AF616748

Molecular Biology Intelligence Unit

Arachidonic Acid in Cell Signaling

Daniele Piomelli, Ph.D.

The Neurosciences Institute
San Diego, California

Springer-Science+Business Media, B.V.

Molecular Biology Intelligence Unit

ARACHIDONIC ACID IN CELL SIGNALING

R.G. LANDES COMPANY
Austin, Texas, U.S.A.

Please address all inquiries to the Publishers:
R.G. Landes Company, 909 Pine Street, Georgetown, Texas, U.S.A. 78626
Phone: 512/ 863 7762; FAX: 512/ 863 0081

DOI 10.1007/978-3-662-05807-7

U.S. and Canada

While the authors, editors and publisher believe that drug selection and dosage and the specifications and usage of equipment and devices, as set forth in this book, are in accord with current recommendations and practice at the time of publication, they make no warranty, expressed or implied, with respect to material described in this book. In view of the ongoing research, equipment development, changes in governmental regulations and the rapid accumulation of information relating to the biomedical sciences, the reader is urged to carefully review and evaluate the information provided herein.

Library of Congress Cataloging-in-Publication Data

Piomelli, Daniele, 1958-
Arachidonic acid in cell signaling / Daniele Piomelli.
p. cm. — (Molecular biology intelligence unit)
Includes bibliographical references and index.
1. Arachidonic acid—Metabolism. 2. Arachidonic acid—Physiological effect. 3. Cellular signal transduction. I. Title. II. Series.
QP752.A7P54 1996
574.19'2477—dc20

96-35506
CIP

www.springer.com/mycopy

Publisher's Note

R.G. Landes Company publishes six book series: *Medical Intelligence Unit, Molecular Biology Intelligence Unit, Neuroscience Intelligence Unit, Tissue Engineering Intelligence Unit, Biotechnology Intelligence Unit* and *Environmental Intelligence Unit.* The authors of our books are acknowledged leaders in their fields and the topics are unique. Almost without exception, no other similar books exist on these topics.

Our goal is to publish books in important and rapidly changing areas of bioscience and environment for sophisticated researchers and clinicians. To achieve this goal, we have accelerated our publishing program to conform to the fast pace in which information grows in bioscience. Most of our books are published within 90 to 120 days of receipt of the manuscript. We would like to thank our readers for their continuing interest and welcome any comments or suggestions they may have for future books.

Shyamali Ghosh
Publications Director
R.G. Landes Company

Dedication

To Livia

CONTENTS

PREFACE

Over the past three decades, the metabolic pathways responsible for the storage, mobilization and metabolism of arachidonic acid in animal tissues have been studied in great detail. Many of the enzymes involved in these biochemical reactions have been characterized, and drugs that interfere with them have been described. Thousands of publications have documented the pharmacological actions exerted by arachidonic acid and its metabolites, collectively known as the eicosanoids, on all systems and organs of the body, as well as their possible physiological roles.

Attempting to give a comprehensive account of such a vast body of knowledge in a single book would be hopeless, and probably useless. This book, instead, has the much more limited purpose of serving as an introductory guide for students or researchers who approach the domain of arachidonic acid signaling from the intersecting fields of cell biology, neurobiology, endocrinology and signal transduction. Thus, I have tried to focus the reader's attention on a small number of findings that, in my opinion, are most apt to illustrate a subject matter. Also, where appropriate, I have described the experimental techniques that have made those findings possible. It goes without saying that, by doing so, I have been obliged to make a series of arbitrary choices, and I apologize to the many researchers whose important contributions I have omitted.

After an Introduction, which places the arachidonic acid cascade in its evolutionary and historical context, the book begins with an analysis of the mechanisms of storage and mobilization of arachidonic acid. Chapter 2 reviews the possible roles of non-esterified arachidonate as an intracellular second messenger. The following chapters describe the enzymes involved in arachidonate metabolism, with special attention to their structure, function and pharmacological inhibition. Chapter 5 briefly outlines the multiple roles of arachidonate metabolites in intracellular and intercellular signaling. Finally, the last chapter reviews the biogenesis and possible functions of anandamide, an arachidonic acid derivative that may act as an endogenous cannabinoid substance in mammalian brains.

I would like to express my gratitude to all my colleagues and friends who have helped me in various ways during the writing of this book and to the Neuroscience Research Foundation, that has provided me with financial support. My special thanks go to Drs. Joseph Gally and Nephi Stella, of The Neurosciences Institute, who read and corrected an initial version of the manuscript; to Prof. Raphael Mechoulam, of the University of Jerusalem, for his comments on chapter 6; to Dr. Atsushi Ichikawa, of Kyoto University, and Dr. John Turk, of the Washington University School of Medicine, who provided me with reprints of their work. But above all, I thank my wife, who endured the many weekends and early mornings devoted to writing.

CHAPTER 1

The Arachidonic Acid Cascade

Arachidonic acid belongs to that relatively small group of chemical substances which were selected in the course of biological evolution to act as informational molecules. These signals are essential for the survival of free-living cells and for coordinating the complex needs of multicellular organisms: as adaptive demands and cell functions changed during evolution, they underwent dramatic transformations in their modes of action, and were called to play diverse roles such as those of pheromones, aggregation factors, hormones, growth factors and second messengers. A classical example of these evolutionary metamorphoses is provided by one of the best-studied signaling molecules, cyclic adenosine monophosphate (cyclic AMP). An ubiquitous intracellular second messenger in metazoa, cyclic AMP acts as an extracellular chemoattractant in the slime mold, *Dictyostelium discoideum*, promoting the aggregation of independent amoebae into worm-like multicellular structures.[1] Like cyclic AMP, arachidonic acid may also have served widely diverse signaling functions during phylogeny.

A simple comparative argument leads to the reasonable, albeit of course untestable, conclusion that the roles of arachidonic acid in cell signaling were established early in evolution. Arachidonic acid and other structurally related fatty acids have been found in virtually all organisms that people have cared to examine. Moreover, these fatty acids appear to be metabolized into an array of biologically active products in most of these organisms, where their products participate in the regulation of a variety of primary functions (e.g., cell aggregation in sponges,[2] osmotic regulation in mollusks,[3,4] wound healing in plants,[5] egg laying in mollusks and echinoderms[6,7]) as well as in a series of adaptations to specific environmental requirements (e.g., defense mechanism in corals,[8,9] suppression of host immune response in parasitic nematodes[10]).

This phylogenetic evidence supports an ancient role of arachidonic acid and its metabolites in the evolution of both intracellular and intercellular communication. It also raises a question which, although intractable from an experimental standpoint, has a certain theoretical interest. What is so special about this fatty acid? Why should arachidonic acid have been selected to play such diverse roles?

We do not know the answer to this question, and probably never will. Yet, two facts allow us to formulate a plausible hypothesis. Arachidonic acid is a polyunsaturated fatty acid (it has 20 carbon atoms and 4 double bonds at positions 5, 8, 11 and 14; its is short-hand designation is therefore 20:4 $\Delta^{5,8,11,14}$), and it is a substantial component of the phospholipids that make up cellular membranes (Fig. 1.1A). Phospholipids containing polyunsaturated fatty acids are thought to provide an environment of proper viscosity for optimal membrane functioning.[10] Thus, it is possible that the selection of arachidonic acid as an informational molecule may derive from nature's tinkering with the lipid composition of early membranous structures. Any enzyme present on the plasma membrane and involved in such tinkering would be ideally placed to act as a signal transducer. For instance, phospholipases (enzymes that mobilize fatty acids from phospholipids) initially employed to adapt, say, arachidonic acid levels in phospholipids to environmental temperatures could have evolved the ability to respond to other external events, resulting in an evolutionary advantage for the organism with this mutation.

A second characteristic of arachidonic acid may have favored its selection as an informational signal over other fatty acids. Signaling molecules must interact with proteins: in other words, they must be able to bind to, and selectively modify the activity of, receptors, enzymes, ion channels, transcription factors, etc. That arachidonic acid may be particularly well-suited to carry out such interactions is suggested by its distinctive low-energy conformations (i.e., the tridimensional shapes that the fatty acid is most

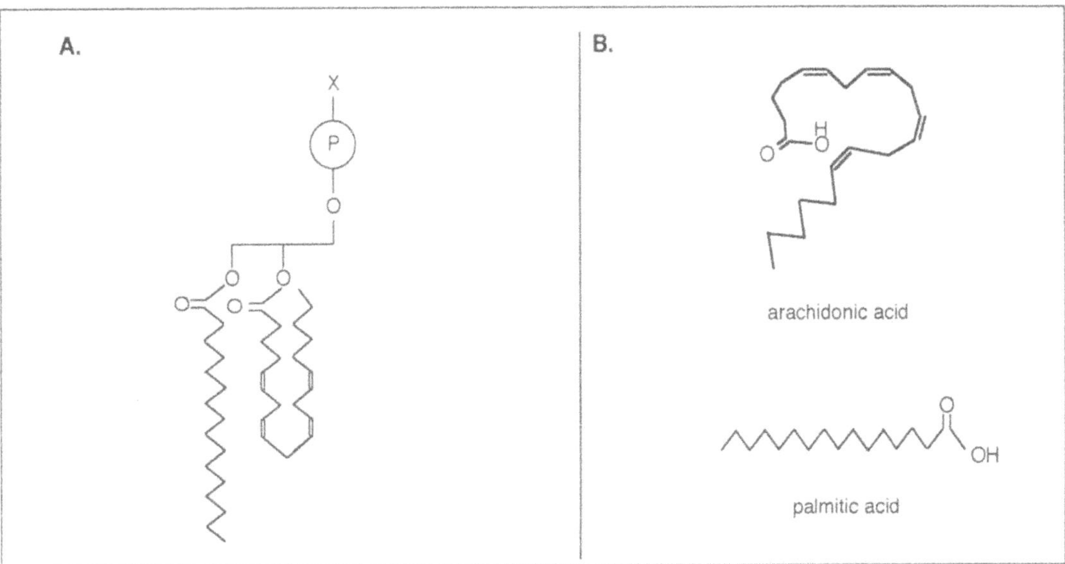

Fig. 1.1. A. Chemical structure of an arachidonate-containing phospholipid. The arachidonate moiety is shown at the sn-2 position of the glycerol backbone, where it occurs most frequently. P indicates a phosphate group, and X one of several possible polar head groups (e.g., choline, ethanolamine, inositol, etc); B. Structure of possible low-energy conformations of arachidonic acid and palmitic acid, as determined by molecular dynamics simulations.[12]

likely to adopt under unrestrained conditions), which have been elucidated by molecular dynamics simulations.[12] After its mobilization from phospholipids, non-esterified arachidonate is thought to adopt the highly curved, hairpin conformation shown in Figure 1.1B. This, in turn, "may directly interact with concave protein surfaces via hydrophobic interactions and/or hydrogen bonding between π-bonds and protein structures."[12] By contrast, saturated fatty acids (e.g., palmitic acid, 16:0) and mono-unsaturated fatty acids (e.g., oleic acid, 18:1 Δ^9) adopt conformations that are linear, or indented in only one point (Fig. 1.1B).[12]

The propensity of non-esterified arachidonate to interact with proteins may also account for the extraordinary richness of its metabolism, witnessed not only by the staggering number of metabolites isolated from animal sources, but also by the fact that a single class of arachidonate metabolites may be produced, in different organisms, through completely distinct enzymatic routes.[13,14]

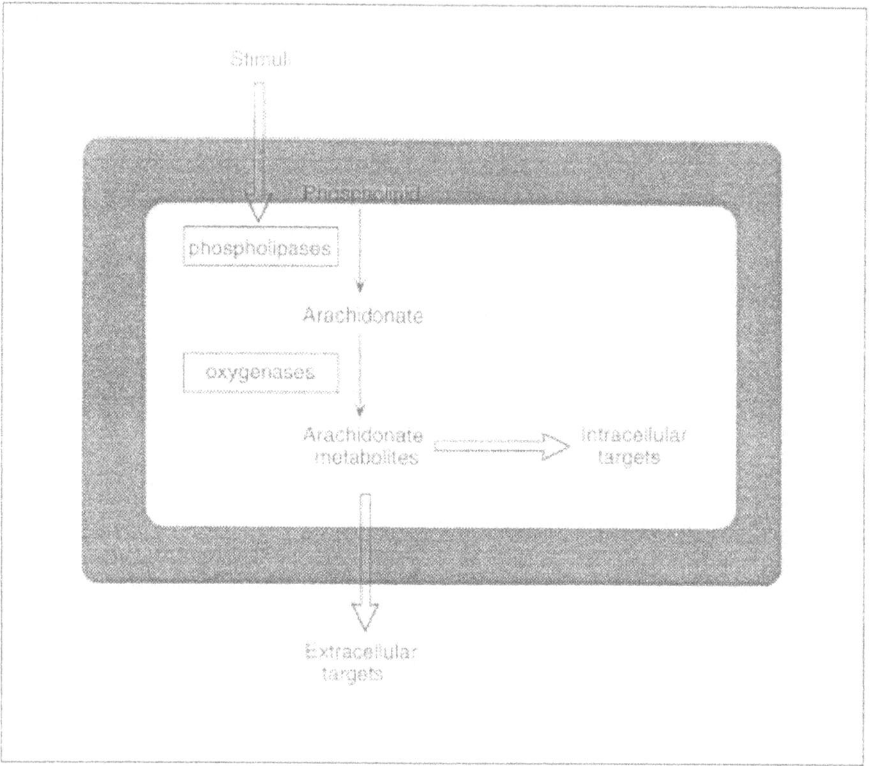

Fig. 1.2. General features of the arachidonate signaling cascade. External stimuli elicit the phospholipase-mediated cleavage of membrane phospholipids, resulting in the mobilization of free arachidonate. Within cells, arachidonate is rapidly metabolized to oxygenated products, which may act either on intracellular target proteins or, after having left the cell of origin, on membrane receptors.

Besides these evolutionary considerations, a closer look at the information available on the phylogeny of the arachidonate cascade may allow us to make some generalizations as to its biogenesis and modes of action, and serve as an introduction to the subject matter of this book. The scheme in Figure 1.2 illustrates some invariant features of the arachidonate cascade based on studies carried out on organisms from a variety of animal phyla.

As we have just seen, arachidonic acid is stored as an ester with glycerol in membrane phospholipids. Signaling begins with the cleavage of these "quiescent" arachidonate stores, an enzymatic reaction driven by external stimuli or by cell-damaging insults (discussed in chapter 2). Arachidonate is mobilized, becoming thus free to interact with protein targets within the cell (chapter 3) or to bind to enzymes that carry out its metabolic transformation into oxygenated derivatives (these are collectively called eicosanoids) (chapter 4). All arachidonate derivatives that serve a signaling function are products of oxygenation, with the one exception of anandamide (N-arachidonoylethanolamine), an endogenous cannabis-like substance. Strictly speaking, anandamide is not an eicosanoid, however, because it is likely produced through a pathway which does not involve the enzymatic transformation of arachidonate, but rather the cleavage of a preformed phospholipid precursor (chapter 6).

Newly formed eicosanoids exert their biological actions either within the cell of origin—by interacting with enzymes, ion channels, etc.—or outside the cell—most commonly by binding to membrane receptors located on neighboring cells (chapter 5). Since they are produced on demand and are inactivated by diffusion or by further metabolism, very small quantities of eicosanoids can usually be found in unstimulated tissues (chapter 4).

In many cases, arachidonate is mobilized from phospholipids together with a few other fatty acids containing multiple double bonds, such as eicosatrienoate (20:3 $\Delta^{8,11,14}$) or linoleate (18:2 $\Delta^{9,12}$). These fatty acids share many of the metabolic routes followed by arachidonic acid, and are also involved in signal transduction. Their biochemical and signaling analogies with arachidonic acid may be well-appreciated by considering the example of linolenic acid. Linolenic acid (18:3 $\Delta^{9,12,15}$) has been extensively studied in plants where, among other things, it serves as precursor for jasmonic acid, a growth-regulating hormone.[15] It has been proposed that jasmonic acid biosynthesis is initiated by pathogens, insect pests or wounding through the production of elicitor signals that interact with specific receptors on the plasma membrane (Fig. 1.3). Activation of elicitor receptors causes the stimulation of a membrane phospholipase activity, which mobilizes linolenate from phospholipids. Non-esterified linolenate is converted then to jasmonate by a series of reactions involving five distinct enzymes: lipoxygenase, allene oxide synthase, allene oxide cyclase, 12-oxo-phytodienoic acid reductase and β-oxidase.[16-18] Finally, jasmonate's pleiotropic effects on plant cells (which include senescence and fruit ripening) are probably triggered by the expression of a number of jasmonate-responsive genes (Fig. 1.3).[15] The analogies of the jasmonate pathway in plants with the arachidonate pathway in vertebrates underline the heuristic value of understanding these lipid signaling cascades in their wider evolutionary context. For more information, the interested reader is referred to a series of review articles published on this subject over the last few years.[19-24]

While our comprehension of the evolutionary history of the arachidonic acid cascade is still very incomplete, we know much more about how the research on the eicosanoid evolved since the discovery of these molecules. Before we turn to the subject of our

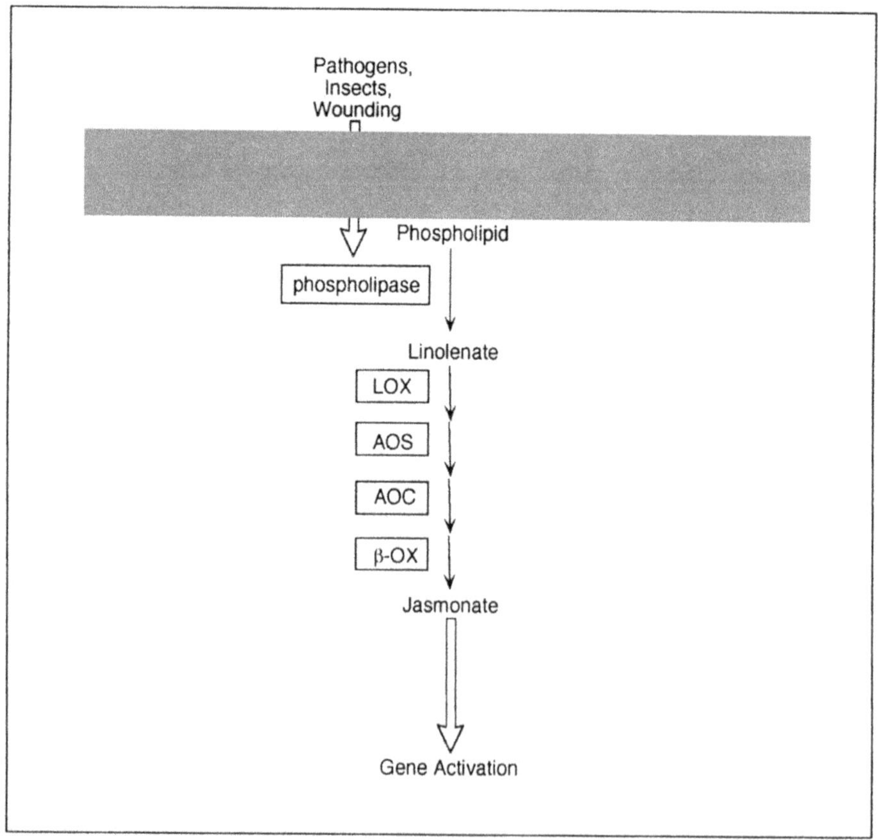

Fig. 1.3. Biosynthesis of the linolenate metabolite and plant growth regulator, jasmonic acid. Abbreviations used are: LOX, lipoxygenase; AOS, allene oxide synthase; AOC allene oxide cyclase; β-OX 12-oxophytodienoic acid reductase and β-oxidase. Based on diagrams published in ref. 15.

next chapter—the biosynthesis, storage and mobilization of arachidonic acid in mammalian cells—it may be worthwhile to set our current thinking within this historical context.

PROSTATE EXTRACTS AND PERFUSED LUNGS

The history of eicosanoid biochemistry and pharmacology began in the mid-1930s, when U.S. von Euler in Sweden and M.W. Goldblatt in the United Kingdom independently reported that mammalian seminal fluid and prostate glands contain a factor that contracts isolated smooth muscle preparations and reduces blood pressure in experimental animals.[25,26] Von Euler, at the Karolinska Institute in Stockholm, was able to extract this factor from prostate glands by using acidified organic solvents, thus demonstrating that its unknown component(s) is an acidic lipid(s). Von Euler named the factor *prostaglandin* but did not attempt to characterize it any further. This task was taken up more than 20 years later by Sune Bergström. With his colleague, Jan Sjöwall, Bergström subjected solvent extracts of sheep prostate glands to a series of chromatographic fractionations by counter-current distribution and paper chromatography. By taking this laborious approach, Bergström and Sjöwall were able to isolate from prostate tissue two compounds in crystalline form, which they called prostaglandin F (PGF, because of its solubility in *ph*osphate buffer) and prostaglandin E (PGE, because of its solubility in *e*ther).[27,28]

The complete structural elucidation of PGE and PGF (renamed later PGE_2 and $PGF_{2\alpha}$) was reported by Bergström in 1962, and revealed that these products are strikingly akin to polyunsaturated fatty acids with 20 carbon atoms and three or four double bonds (Fig. 1.4). It was logical therefore to suppose that arachidonic acid could serve as a common precursor for this family of bioactive lipids. This hypothesis was substantiated in Bergström's laboratory and, independently, by van Dorp and his coworkers in Holland. Using ram seminal vesicles, both groups demonstrated that exogenous arachidonic acid can be converted into the prostaglandins, PGE_2 and $PGF_{2\alpha}$, by a novel enzyme activity, now known as cyclooxygenase or prostaglandin H synthase (for review, see ref. 29). But PGE_2 and $PGF_{2\alpha}$ were only the beginning.

A few years after the discovery of cyclooxygenase, Priscilla J. Piper and John R. Vane, then at the Royal College of Surgeons in London, reported that during anaphylaxis the lungs of sensitized guinea-pigs release a smooth-muscle contracting factor, which they called rabbit aorta contracting substance (RCS). Thanks to their remarkable command over the technique of biological assay, which was at that time the pharmacologist's main experimental asset, Piper and Vane were able to gather a great deal of information on the nature of RCS. Most importantly, they established that RCS is a very short-lived, novel substance whose release is prevented by anti-inflammatory drugs like aspirin or indomethacin.[30] These results set the stage for two essential advances: first, the identification of

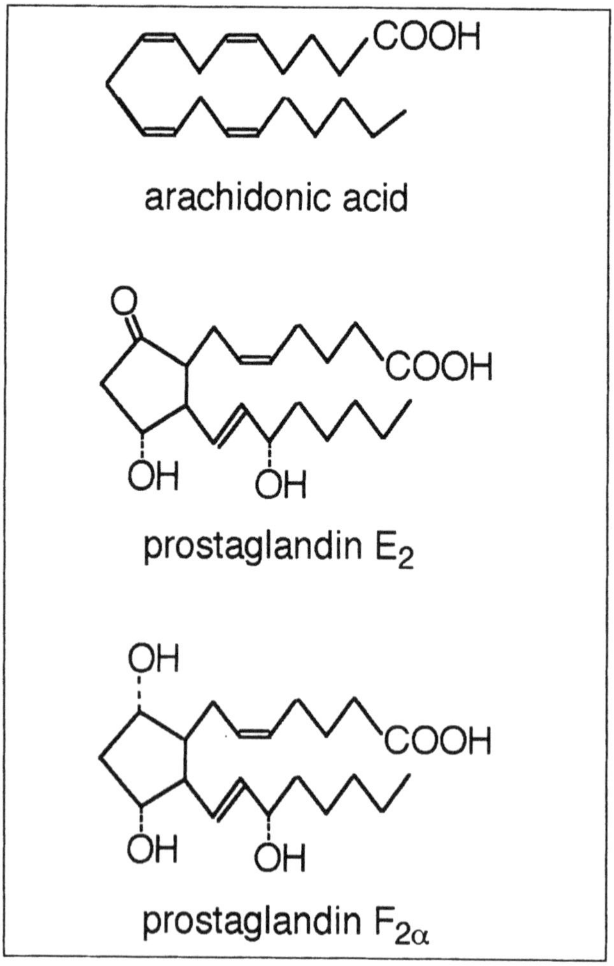

Fig. 1.4. Chemical structures of prostaglandin E_2 and prostaglandin $F_{2\alpha}$, showing their structural relationship with arachidonic acid.

RCS as an unstable cyclooxygenase metabolite of arachidonic acid, thromboxane A_2 (TXA_2);[31] second, the finding that aspirin and other non-steroidal anti-inflammatory drugs exert their effects by inhibiting cyclooxygenase activity—a discovery of momentous therapeutic importance.[32,33]

TXA_2 is an extremely potent vasoconstrictory and platelet-aggregating agent, and is produced in large quantities by stimulated platelets. While these properties of TXA_2 could well account for the anti-coagulating effect of aspirin, they also raised an intriguing question: how can the blood vessel cope with the continuous production by platelets of such a powerful constricting and thrombotic substance? Salvador Moncada, John R. Vane and their collaborators hypothesized that cells in the blood vessel wall release an anti-thrombotic and vasorelaxant compound that antagonizes the effects of TXA_2. When they tested this possibility, they did not only discover that such a compound exists, but also that it is, like TXA_2, a short-lived cyclooxygenase product.[34] The ensuing chemical characterization of this metabolite, which they named prostacyclin (PGI_2), revealed its relationship with 6-keto $PGF_{1\alpha}$—PGI_2's stable hydrolysis product isolated a few years earlier by Cecil R. Pace-Asciak and Leonard S. Wolfe in Toronto.[35]

Guinea pig lungs, from which TXA_2 was first isolated, had not yet ceased to surprise. Pharmacologists had known since many years that, when challenged with an antigen, sensitized guinea-pig lungs generate a factor with slow-developing smooth-muscle contracting properties, the slow-reacting substance of anaphylaxis (SRS-A).[36] Like von Euler's prostate extracts, SRS-A was known to be an acidic lipid, but its biological actions could not be mimicked by any known prostaglandin, and its formation was not inhibited by aspirin. What kind of lipid substance was SRS-A, then?

An initial series of studies provided evidence that SRS-A was likely to be an arachidonic acid metabolite produced by a biochemical pathway distinct from cyclooxygenase. This evidence was based on chemically impure preparations of SRS-A, however, and so remained circumstantial until 1979, when the coupling of a relatively new analytical technique (high-performance liquid chromatography, HPLC) with classical bioassay methods allowed Robert C. Murphy, Bengt Samuelsson and their collaborators at the

Karolinska Institute in Stockholm to purify SRS-A from a mouse mastocytoma cell line in sufficient quantities to carry out its chemical characterization. These researchers capitalized also on the discoveries, by D.H. Nugteren in 1974 and by Pierre Borgeat and Bengt Samuelsson in 1979, that arachidonic acid is a substrate not only for cyclooxygenase but also for a variety of lipoxygenase activities.[37,38] With this background information at their disposal, Murphy and coworkers were able to demonstrate unambiguously that the active principle in SRS-A is an arachidonic acid metabolite produced through the newly-described 5-lipoxygenase pathway.[39] They called this metabolite leukotriene C_4 (LTC_4) after its cellular source and the presence in its chemical structure of a conjugated triene, which confers to it a typical absorbance spectrum in the ultraviolet.

Even a short outline of the history of the eicosanoids, such as the present one, would be seriously incomplete without mentioning how the third, main pathway of arachidonic acid metabolism, cytochrome P_{450}, came to be discovered. This occurred through a series of experiments that differed in their logical unfolding from those that led to the discovery of prostaglandins and leukotrienes. The latter were identified first as biologically active factors, distinguished from other known biological substances by their peculiar pharmacological properties (selective effects on smooth muscle activity, platelet aggregation, etc.). The brilliant biochemical analyses that led to the structural characterization of these eicosanoids came only afterwards. In a way, it was pharmacology that led the way to biochemistry.

The opposite is true for arachidonic acid metabolism via the cytochrome P_{450} pathway. Cytochrome P_{450} had been recognized for many years as a component of the liver's microsomal 'detoxifying' enzyme system. Before the 1980s, cytochrome P_{450} was thought to act mainly on xenobiotics, particularly on drugs, and to have evolved in animals, not to participate in signal transduction, but as the result of their exposure to toxic alkaloid in plants.

Thus, when J. Capdevila, J.R. Falck, E.H. Oliw and their collaborators reported that liver cytochrome P_{450} converts arachidonic acid into a family of novel epoxides,[40,41] researchers in the eicosanoid field looked at these studies with concealed skepticism. Nobody argued, of course, against the biochemical validity of these observations. But, for lack of biological actions, some regarded them as physiologically irrelevant. It was mainly through the efforts of N.R. Ferreri, J.C. McGiff and coworkers, at the New York Medical College in Valhalla, that the crucial involvement of cytochrome P_{450} eicosanoids in the control of ion fluxes in kidney and corneal tissues was first suggested and then established, leading to a recognition of their biological importance (for review, see ref. 42). A similar approach, moving from biochemistry to pharmacology, was successfully adopted with other families of arachidonic acid metabolites that intervene in cell signaling, including the hepoxilins[43,44] (12-lipoxygenase metabolites discovered by Cecil R. Pace-Asciak) and the lipoxins (metabolites of multiple lipoxygenase activities discovered by Charles N. Serhan in Bengt Samuelsson's laboratory).[45]

This cursory historical overview of the research on the eicosanoids has no pretension of completeness, and I apologize to the many investigators in this field whose important contributions I have omitted. Several such contributions will be acknowledged in following chapters. Here, I have primarily attempted to underline a theme recurring in the discovery of the arachidonate signaling cascade as well as in other events in the history of pharmacology—such as the discovery of the endothelium-derived relaxant factor (later identified as nitric oxide) in the 1980s[46] and of the endogenous cannabinoid substance, anandamide, in 1992.[47] Even in the era of molecular biology, receptor cloning and 'reverse pharmacology', major pharmacological advances can still be made by linking imaginative inductions, based on the results of simple bioassay systems, to rigorous biochemical analyses.

References

1. Gerisch G. Chemotaxis in *Dictyostelium*. Annu Rev Physiol 1982; 44:535-552.
2. Rich AM, Weissman G, Anderson C, et al. Calcium-dependent aggregation of marine sponge cells is provoked by leukotriene B_4 and inhibited by inhibitors of arachidonic acid oxidation. Biochem Biophys Res Commun 1984; 121:863-870.
3. Graves SY, Dietz TH. Prostaglandin E_2 inhibition of sodium transport in the freshwater mussel. J Exp Biol 1979; 210:195-201.

4. Freas W, Grollman. Uptake and binding of prostaglandins in the marine bivalve, *Modiulus demissus*. J Exp Zool 1981; 216: 225-233.
5. Bell E, Creelman RA, Mullet JE. A chloroplast lipoxygenase is required for wound-induced jasmonic acid accumulation in *Arabidopsis*. Proc Natl Acad Sci USA 1995; 92:8675-8679.
6. Hill EM, Holland DL. Identification and egg hatching activity of monohydroxy fatty acid eicosanoids in the barnacle *Balanus balanoides*. Proc R Soc Lond 1992; 247:41-46.
7. Holland DL, East J, Gibson H, Clayton E, Oldfield A. Identification of the hatching factor of the barnacle *Balanus balanoides* as the novel eicosanoid 10,11,12 trihydroxy-5,8,14,17-eicosatetraenoic acid. Prostaglandins 1985; 29:1021-1029.
8. Brash AR, Baertschi SW, Ingram CD, Harris TM. On non-cyclooxygenase prostaglandin synthesis in the sea whip coral, *Plexaura homomalla*: an 8(R)-lipoxygenase pathway leads to formation of an alpha-ketol and a racemic prostanoid. J Biol Chem 1987; 262:15829-15839.
9. Corey EJ, Matsuda SPT, Nagata R, Cleaver MB. Biosynthesis of 8(R)-HPETE and preclavulone A from arachidonate in several species of caribbean coral. A widespread route to marine prostanoids. Tetrahedron Lett 1988; 29:2555-2558.
10. Salfsky B, Wang Y-S, Fusco AC, Antonacci J. The role of essential fatty acids and prostaglandins in cercarial penetration (*Schistosoma mansoni*). J Parasitol 1984; 70:656-660.
11. Thompson Jr GA. The regulation of membrane lipid metabolism. CRC Press, Boca Raton, 1992.
12. Rich MR. Conformational analysis of arachidonic acid and related fatty acids using molecular dynamics simulations. Biochim Biophys Acta 1993; 1178:87-96.
13. Song WC, Brash AR. Investigation of the allene oxide pathway in the coral *Plexaura homomalla*: formation of novel ketols and isomers of prostaglandin A_2 from 15-hydroxyeicosatetraenoic acid. Arch Biochem Biophys 1991; 290:427-435.
14. Brash AR, Baertschi SW, Harris TM. Formation of prostaglandin A analogues via an allene oxide. J Biol Chem 1990; 265:6705-6712.
15. Creelman RA, Mullet JE. Jasmonic acid distribution and action in plants: regulation during development and response to biotic and abiotic stress. Proc Natl Acad Sci USA 1995; 92:4114-4119.
16. Ryan CA. The search for the proteinase inhibitor-inducing factor. Plant Mol Biol 1992; 19:123-133.
17. Hamberg M, Gardner HW. Oxylipin pathway to jasmonates: biochemistry and biological significance. Biochim Biophys Acta 1992; 1165:1-18.

18. Song W-C, Brash AR. Purification of an allene oxide synthase and identification of the enzyme as a cytochrome P-450. Science 1991; 253:781-784.
19. Stanley-Samuelsson DW, Loher W. Evolutionary aspects of prostaglandins and other eicosanoids in invertebrates. Wiley-Liss, 1990.
20. Gerwick WH, Nagle DG, Proteau PJ. Oxylipins from marine invertebrates. In: Scheuer PJ, ed. Topics in Current Chemistry. Berlin: Springer Verlag, 1993.
21. Gerwick WH. Carbocyclic oxylipins of marine origins. Chem Rev 1993; 93:1807-1823.
22. Stanley-Samuelsson DW. Physiological roles of prostaglandins and other eicosanoids in invertebrates. Biol Bull 1987; 173:p2-109.
23. Stanley-Samuelsson DW. Comparative eicosanoid physiology in invertebrate animals. Am J Physiol 1991; 29:R849-R853.
24. Bundy GL. Nonmammalian sources of eicosanoids. Adv Prostaglandin Thromboxane Leukotriene Res 1985; 14:229-262.
25. von Euler US. On the specific vasodilating and plain muscle stimulating substance from accessory genital glands in man and certain animals (prostaglandin and vesiglandin). J Physiol 1936; 88:213-234.
26. Goldblatt MW. Properties of human seminal fluid. J Physiol 1935; 84:208-218.
27. Bergström S, Sjöwall J. The isolation of prostaglandin F from sheep prostate gland. Acta Chem Scand 1960; 14:1693-1700.
28. Bergström S, Sjöwall J. The isolation of prostaglandin E from sheep prostate gland. Acta Chem Scand 1960; 14:1701-1705.
29. Bergström S, Samuelsson B. The prostaglandins. Endevour 1968; 27:109-113.
30. Piper PJ, Vane JR. Release of additional factors in anaphylaxis and its antagonism by anti-inflammatory drugs. Nature 1969; 223:29-35.
31. Hamberg M, Svensson J, Samuelsson B. Thromboxanes: a new group of biologically active compounds derived from prostaglandin endoperoxides. Proc Natl Acad Sci USA 1975; 72:2994-2998.
32. Ferreira SH, Moncada S, Vane JR. Indomethacin and aspirin abolish prostaglandin release from the spleen. Nature [New Biology] 1971; 231:237-239.
33. Smith JB, Willis AL. Aspirin selectively inhibits prostaglandin production in human platelets. Nature [New Biology] 1971; 231:235-237.
34. Moncada S, Gryglewski R, Bunting S, Vane JR. An enzyme isolated from arteries transforms prostaglandin endoperoxides to an unstable substance that inhibits platelet aggregation. Nature 1976; 263:663-665.
35. Pace-Asciak CR, Wolfe LS. A novel prostaglandin derivative formed from arachidonic acid by rat stomach homogenates. Biochemistry 1971; 10:3657-3664.

36. Feldberg W, Kellaway CH. Liberation of histamine and formation of lysolecithin-like substances by cobra venom. J Physiol 1938; 94:187-226.
37. Borgeat P, Samuelsson B. Metabolism of arachidonic acid in polymorphonuclear leukocytes. Structural analysis of novel hydroxylated compounds. J Biol Chem 1979; 254:7865-7869.
38. Nugteren DH. Arachidonate lipoxygenase in blood platelets. Biochim Biophys Acta 1975; 380:299-307.
39. Murphy RC, Hammarström S, Samuelsson B. Leukotriene C: A slow-reacting substance from murine mastocytoma cells. Proc Natl Acad Sci USA 1979; 76:4275-4279.
40. Capdevila J, Chacos N, Werringloer J, Prough RA, Estabrook RW. Liver microsomal cytochrome P450 and the oxidative metabolism of arachidonic acid. Proc Natl Acad USA 1981; 78:5362-5366.
41. Oliw EH, Guengerich FP, Oates JA. Oxygenation of arachidonic acid by hepatic monooxygenases. J Biol Chem 1982; 257: 3771-3781.
42. McGiff JC. Cytochrome P450 metabolism of arachidonic acid. Annu Rev Pharmacol Toxicol 1991; 31:339-369.
43. Pace-Asciak CR, Granström E, Samuelsson B. Arachidonic acid epoxides. Isolation and structure of two hydroxyepoxide intermediates in the formation of 8,11,12 and 10,11,12-trihydroxyeicosatrienoic acids. J Biol Chem 1983; 258:6835-6840.
44. Piomelli D, Shapiro E, Zipkin R, Schwartz JH, Feinmark SJ. Formation and action of 8-hydroxy-11,12-epoxy-icosatrienoic acid in *Aplysia*: a possible second messenger in neurons. Proc Natl Acad Sci USA 1989; 86:1721-1725.
45. Serhan CN. Lipoxin biosynthesis and its impact in inflammatory and vascular events. Biochim Biophys Acta 1994; 1212:1-25.
46. Furchgott RF, Zawadski JV. The obligatory role of endothelial cells in the relaxation of arterial smooth muscle by acetylcholine. Nature 1980; 288:373-376.
47. Devane WA, Hanus L, Breuer A et al. Isolation and structure of a brain constituent that binds to the cannabinoid receptor. Science 1992; 258:1946-1949.

CHAPTER 2

Biosynthesis, Storage and Mobilization of Arachidonic Acid

A basic principle that has emerged from the last decade of research on lipid signaling is that membrane phospholipids should be regarded not only as structural components of the cell, but also as precursors for transmembrane, or even transcellular, signaling molecules. In Table 2.1, I have listed a series of examples to illustrate this principle. Three components play a major role in

Table 2.1. Common themes in lipid signaling

Precursor	phospholipase	signaling molecule(s)	biological effectors
PI, PC	phospholipase C	1,2 diacylglycerol inositol-trisphosphate	protein kinase C intracellular [Ca2+]
sphingomyelin	sphingomyelinase	ceramide	ceramide-activated protein kinase and phosphatase
NAPE	phospholipase D	N-acylethanolamines	cannabinoid receptors
PC	phospholipase D	phosphatidic acid	phosphatidic acid-dependent protein kinase
phosphatidic acid	phospholipase A_2	lysophosphatidic acid	lyso-phosphatidic acid receptors
PI, PC, PE	phospholipase A_2	arachidonate	protein kinases, ion channels, further metabolism

Abbreviations: PC, phosphatidylcholine; PE, phosphatidylethanolamine; PI; phosphatidylinositol; NAPE, N-acyl phosphatidylethanolamine.

each of these examples: a phospholipid precursor, often but not exclusively found in the plasma membrane; a set of enzymes that synthesize this precursor; and a stimulus-activated phospholipase that hydrolyzes it, producing one or more metabolites with distinct biological effects.

Cell-specificity and stimulus-specificity are provided for by variations on any one of these aspects. For instance, formation of ceramide, a lipid messenger involved in programmed cell death,[1] depends on the presence in a given cell type of both a complement of sphingomyelin-synthesizing enzymes and a receptor-activated sphingomyelinase (a phospholipase that cleaves sphingomyelin producing ceramide). Likewise, the cell-specific actions of ceramide depend on the presence of appropriate effectors (e.g., ceramide-activated protein kinase or protein phosphatase) as well as their protein substrates.[1]

For all their common elements, the various avatars of phospholipid hydrolysis are different enough to call for a great deal of individual attention. In the case of arachidonic acid, three processes set the stage for its participation in cellular signaling: biosynthesis from linoleic acid, incorporation into phospholipids, and mobilization from phospholipids. To these processes, we shall now direct our attention.

ARACHIDONIC ACID BIOSYNTHESIS

The enzymatic reactions involved in the biosynthesis of arachidonic acid in mammalian cells are shown in Figure 2.1. As in the case with other polyunsaturated fatty acids, they consist of alternating steps of chain elongation (i.e., addition of an acetyl unit to the carboxyl-terminal end) and desaturation (i.e., formation of a new double bond).[2] In mammals, linoleic acid is the ultimate precursor for arachidonic acid: an 'essential' fatty acid, linoleic acid cannot be synthesized and must be obtained from plants, in which it is usually very abundant.

Arachidonic acid biosynthesis does not occur in all cell types in the mammals' body. Neurons, for instance, can neither elongate nor desaturate linoleic acid and must rely therefore on other cells for their supply of free arachidonate. What type of cells? Three complementary sources have been identified: hepatocytes, which

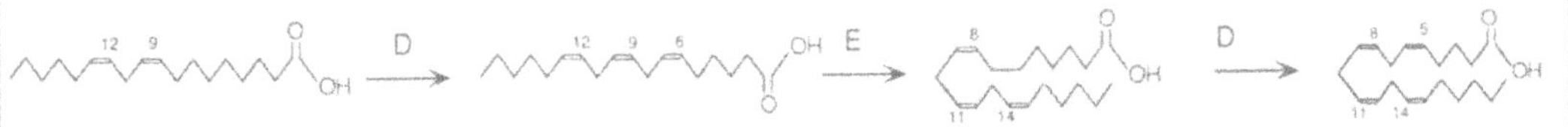

Fig. 2.1. Biosynthesis of arachidonic acid from linoleic acid in animal tissues. Linoleic acid (short-hand designation, 18:2 $\Delta^{9,12}$), derived from the diet, is converted to γ-linolenic acid (18:3 $\Delta^{6,9,12}$) by a desaturase activity (D). Two subsequent steps of elongation (E) and desaturation are necessary to complete arachidonic acid biosynthesis.

release arachidonate into the bloodstream, where the fatty acid is transported either as a complex with serum albumin or packaged as a triacylglycerol ester in the hydrophobic core of plasma lipoproteins; cerebral microvascular endothelium, which constitutes the blood-brain barrier; and astroglia.[3,4] Thus, depending on the cell type, arachidonic acid may be either synthesized de novo from linoleic acid or obtained from extrinsic sources.

INCORPORATION INTO PHOSPHOLIPIDS

Whether newly synthesized or not, arachidonate is taken up by cells and stored in phospholipids so rapidly that, under resting conditions, only trace amounts of it may be found in free form.[5] This is not usually the case with other long-chain fatty acids, such as palmitic, stearic and oleic acids, which are incorporated into phospholipids at much lower rates and can accumulate within cells, giving rise to a significant non-esterified pool.[6]

This tight and selective control on the concentration of intracellular free arachidonate is justified by the signaling role of this lipid, and is exerted by two concerted enzymatic activities: arachidonoyl-coenzyme A (CoA) synthetase and arachidonoyl-CoA:lysophospholipid transferase. The reactions catalyzed by these two enzymes are illustrated in Figure 2.2.

In the presence of reduced CoA (CoA-SH), ATP and Mg^{2+}, arachidonoyl-CoA synthetase converts arachidonate into arachidonoyl-CoA, a process referred to as 'activation'. Arachidonoyl-CoA synthetase is quite specific for arachidonic acid and for 8,11,14-eicosatrienoic acid; saturated and monosaturated long-chain fatty acids are, instead, activated by a distinct long-chain acyl-CoA synthetase activity with broader substrate specificity. Some properties of the platelet synthetase, studied by the group of Philip W. Majerus (at the Washington University School of Medicine in Saint Louis), are reported in Table 2.2.[7,8]

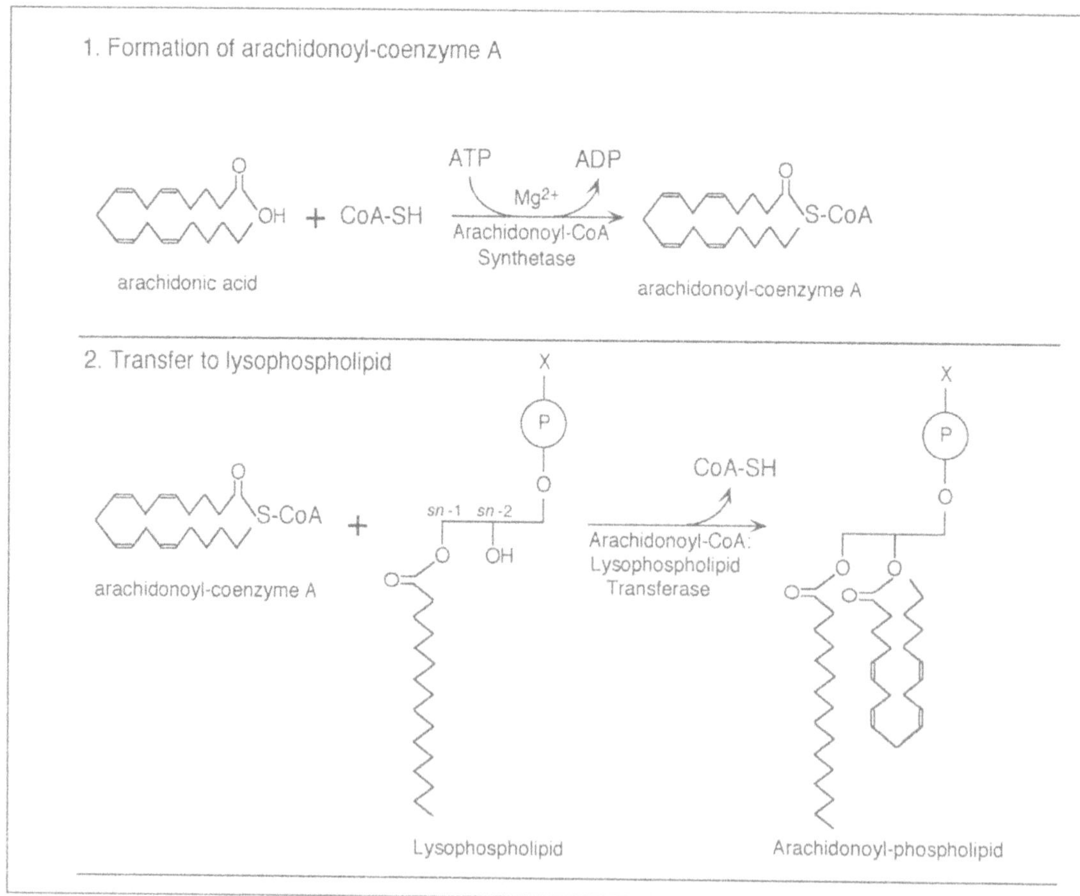

Fig. 2.2. Esterification of arachidonic acid to membrane phospholipids. (1) ATP-dependent condensation of free arachidonate with coenzyme A (CoA-SH), catalyzed by arachidonoyl-CoA synthetase; (2) Transfer of the arachidonoyl moiety from arachidonoyl-CoA to lysophospholipid, producing sn-2 arachidonoyl phospholipid.

Table 2.2 Requirements for arachidonoyl-CoA synthetase activity in human platelets

Cofactor	Arachidonoyl-CoA
Complete system	100%
w/out ATP	0
w/out CoA-SH	0
w/out Mg2+	2.5%

Like non-esterified arachidonate, arachidonoyl-CoA is very difficult to isolate from tissues because the subsequent step of esterification to phospholipids proceeds with high efficiency. This reaction, catalyzed by arachidonoyl-CoA:lysophospholipid transferase, consists in the intermolecular transfer of the fatty acyl moiety of arachidonoyl-CoA to the *sn*-2 position of lysophospholipid, yielding *sn*-2 arachidonoyl-phospholipid (Fig. 2.2).[9] The cellular concentration of lysophospholipid is also very low under normal conditions, suggesting that the levels of this intermediate are subject to some sort of regulatory control. That such control may exist is suggested by evidence reviewed later in this chapter, in the section on calcium-independent phospholipases A_2.

While storage in membrane phospholipids is, from the standpoint of signal transduction, a predominant fate of arachidonoyl-CoA, additional pathways do exist. For instance, as shown in Figure 2.3, arachidonoyl-CoA may be converted back to arachidonate by a fatty acyl-CoA hydrolase activity, whose biological roles, if any, remain unknown.[10] Moreover, arachidonoyl-CoA is an acceptable substrate for other acyltransferase activities, and it may therefore be used for the synthesis of cholesterol esters (a reserve supply of cholesterol) and triacylglycerols.[10] Arachidonic acid in

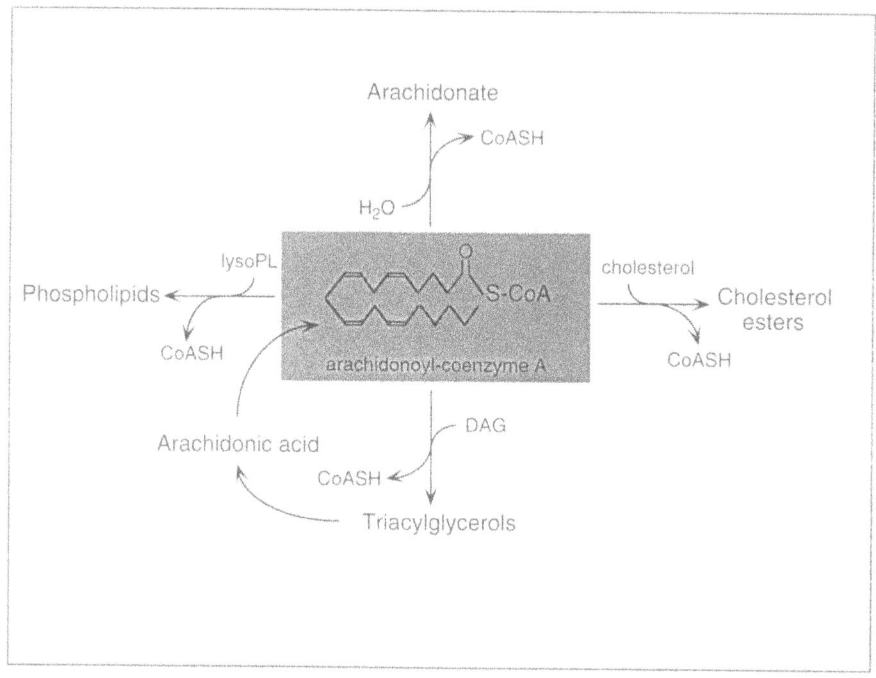

Fig. 2.3. Overview of arachidonoyl-CoA metabolism in mammalian cells. Four major metabolic routes are illustrated: (1) formation of phospholipids; (2) formation of triacylglycerols; (3) formation of cholesteryl esters; (4) hydrolysis. Arachidonate stored in triacylglycerol may become available for incorporation into phospholipid through the action of the enzyme triacylglycerol lipase.

triacylglycerols may be utilized for energy production through the β-oxidation cycle. The triacylglycerol arachidonoyl moiety may also be mobilized by a lipase activity, converted to arachidonoyl-CoA, and transferred to phospholipid.[11] The occurrence in cells of this sequence of reactions implies that triacylglycerols may serve as a 'secondary' reservoir of arachidonate (Fig. 2.3).

PATHWAYS OF ARACHIDONATE MOBILIZATION

We have seen how arachidonic acid may be synthesized and stored in cells. We turn now to the more complex subject of its stimulus-dependent mobilization from membrane phospholipids.

A large number of primary messengers (e.g., hormones, autacoids, neurotransmitters, and growth factors) can control arachidonate mobilization. As a rule, they do so in a seemingly straightforward way, i.e., by activating membrane receptors linked positively or negatively to phospholipases. But the structures of these receptors, the phospholipases they are linked to and their precise coupling mechanisms may differ widely from cell to cell and even within the same cell type. In the following pages, I will describe various types of enzyme activities involved in arachidonate mobilization, the reactions they catalyze, their molecular structures (when known) and their pharmacology. I will also examine the mechanisms underlying the receptor-dependent regulation of these enzyme activities. Finally, I will illustrate some examples of second messenger cross-talk that implicate the arachidonic acid cascade.

Before starting with our first topic, the phospholipases, a few words of introduction may be helpful. We shall encounter three classes of phospholipases that participate in the formation of free arachidonate: phospholipase A_2 (PLA_2), phospholipase C (PLC) and phospholipase D (PLD). Their sites of attack on phospholipid are shown in Figure 2.4. PLA_2s catalyze the hydrolysis of the fatty ester bond at the *sn*-2 position of phospholipid and can therefore mobilize arachidonate in a single-step reaction. We shall focus first on this family of enzymes. In contrast with PLA_2s, PLCs and PLDs do not release arachidonate directly. Rather, they generate lipid products containing arachidonate (diacylglycerol and phosphatidic acid, respectively), from which the fatty acid can be subsequently released by PLA_2, diacylglycerol lipase or monoacylglycerol lipase activities. We shall look at PLCs and PLDs only from the standpoint of their role in arachidonate mobilization, as their other important functions in cellular signaling are beyond the scope of this book.

PHOSPHOLIPASES A_2 : AN OVERVIEW

PLA_2s constitute a large group of acylhydrolases that, as we have just seen, catalyze the hydrolysis of the *sn*-2 fatty ester bond of phospholipid, forming free fatty acid and lysophospholipid (Fig. 2.4). They participate in a variety of biological processes—from digestion of food to membrane remodeling, from host defense to signal transduction. As one may expect from such diverse roles, PLA_2s are also widely divergent with regard to their protein structures, gene sequences, cellular localizations, activation mechanisms and pharmacological inhibition (for review, see refs. 12-17).

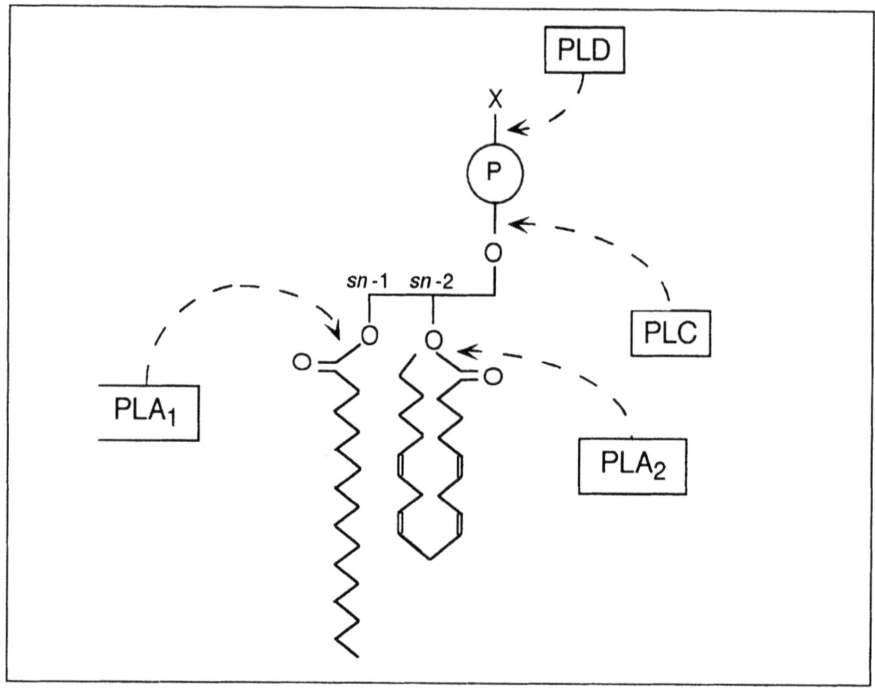

Fig. 2.4. Sites of attack of various phospholipases (PL) on phospholipid. PLB activity, not shown in figure, catalyzes the hydrolysis of ester bonds at both the sn-1 *and* sn-2 *positions of the glycerophospholipid backbone.*

A popular classification distinguishes between extracellular, or low molecular mass (14-18 kDa) PLA_2s—which are referred to as 'secreted', or sPLA_2s—and intracellular or high molecular mass (31-110 kDa) PLA_2s—'cytosolic', or cPLA_2s. Although helpful in its succinctness, this classification suffers from being somewhat oversimplified, and an alternative one, summarized in Table 2.3, has been proposed.[17] It postulates the existence of at least four groups of PLA_2s identified on the basis of their homologies in amino acid sequence.

The enzymes belonging to groups I, II and III share a series of common structural and mechanistic features: they all have low molecular masses (14-18 kDa), a high disulfide bond content, and require relatively high concentrations of calcium ions (in the mM range) for activation in vitro. While these three PLA_2 groups correspond to the sPLA_2s of traditional classifications, it is important to underline that at least two of them, groups I and II, are found in mammalian tissues as both secreted and non-secreted, membrane-associated enzymes. As we will see below, their mechanism of activation may be markedly different.

Group IV comprises a PLA_2 with high molecular mass (85 kDa) which preferentially hydrolyzes phospholipids containing arachidonic acid at the *sn*-2 position and requires for activation concentrations of calcium ions in the high nanomolar or low micromolar range. Group IV PLA_2 corresponds only partially to the enzymes designed in the literature as $cPLA_2$s: in fact, the latter term is used sometimes to encompass also other cytosolic PLA_2s (calcium-dependent as well as calcium-independent) which have not been characterized at the molecular level and therefore have not been assigned yet to any specific group. A distinctive feature of group IV PLA_2—the specificity for arachidonate-containing phospholipids—has attracted a great deal of research over the last several years. This attention is well deserved, as we shall see in the next section.

Table 2.3. Classification of phospholipases A_2

Source	**Location (kD)**	**Molecular Mass Stimulation**	**Calcium**
Group I. A. Cobra venom B. Mammalian Pancreas	Secreted, cell-associated	13-15	mM
Group II. A. Viper venom; human synovial fluid, platelets B. Gaboon viper venom	Secreted, cell-associated	13-15	mM
Group III. Bee venom	Secreted	16-18	mM
Group IV. Various mammalian tissues	Cytosolic	85	mM

This classification is not exhaustive. Because of limited structural information, other phospholipases A_2 have not yet been classified.

GROUP IV PHOSPHOLIPASE A_2

The purification of group IV, or cytosolic, PLA_2, which had proved particularly difficult because of the low levels of enzyme present in tissues, was reported in 1990 and 1991 simultaneously by several laboratories.[18-21] These found that purified group IV PLA_2 has an apparent molecular mass of 110 kDa, when determined by sodium dodecylsulfate polyacrylamide gel electrophoresis (SDS-PAGE), and an isoelectric point of 5.1. They also found that group IV PLA_2 is maximally activated by concentrations of calcium ions between 0.3 μM and 1 μM, shows a high preference for phospholipid substrates containing arachidonic acid at the *sn*-2 position and is present in a variety of mammalian tissues (as determined, for instance, by immunochemical analysis).[22]

The molecular cloning of group IV PLA_2 revealed a predicted protein sequence of 749 amino acids, corresponding to an actual molecular mass of 85 kDa, and confirmed two characteristics of this enzyme that investigators had previously suspected on the weight of indirect evidence. First, that group IV PLA_2 bears no sequence homology with group I-III PLA_2s; second, that the predicted sequence of group IV PLA_2 contains a 45 amino acid region homologous to those seen in proteins that translocate from cytosol to plasma membrane in response to physiological increments in intracellular free calcium (for instance, protein kinase C and phosphoinositide-specific PLC).[23,24] Indeed, purified group IV PLA_2 was shown to associate with membrane vesicles when exposed, in vitro, to concentrations of calcium of 0.3 μM or higher (maximal at 1 μM).[18] Further studies demonstrated that calcium-dependent membrane association is a central step, but not the only one, in the sequence of molecular events that lead to the activation of this PLA_2. In addition to membrane association, protein phosphorylation is also required.

In vitro, purified group IV PLA_2 is a substrate for at least three major protein kinases: protein kinase C (PKC), cyclic AMP-dependent protein kinase (PKA) and mitogen-activated protein kinase (MAP kinase). MAP kinase phosphorylates group IV PLA_2 on a serine residue (serine 505), causing a dramatic increase in the enzyme's catalytic activity; the sites phosphorylated by PKC or PKA are different from the MAP kinase site and these phosphorylations do not appear to markedly affect the catalytic activity[25] (see ref. 26 for contrasting results). Yet, stimulating PKC in intact cells with drugs (e.g., phorbol esters) or with receptor agonists, typically produces an increase in PLA_2 activity accompanied by arachidonate mobilization. How may we explain this incongruity? Evidence indicates that the stimulating effect of PKC on group IV PLA_2 may be indirect, and mediated by MAP kinase. For instance, it was shown that phorbol esters-induced phosphorylation of group IV PLA_2 takes place not on the PKC sites, but on serine 505, the MAP kinase site.[25]

Thus, calcium-dependent membrane translocation and MAP kinase phosphorylation may act synergistically to activate group IV PLA_2. The simplified model depicted in Figure 2.5 summarizes these interactions: transmembrane receptors coupled to PLC activation (e.g., purinergic receptors on fibroblasts) evoke both intracellular calcium rises and PKC activation. The former may cause PLA_2 to translocate from the cytosol to the membrane, where its phospholipid substrate is located. The latter may initiate in turn a chain of reactions ending in MAP kinase activation and PLA_2 phosphorylation. Although the convergence of these two events may be necessary for PLA_2 activity, it may still be insufficient for its maximal expression; the role of transducing G proteins will be considered later in this chapter.

Tyrosine-kinase receptors (e.g., receptors for platelet-derived growth factor, PDGF, or endothelium growth factor, EGF) are also linked to group IV PLA_2 activation: Ras-related low molecular-mass GTPases (e.g., Ras itself or Rac) are likely to participate in coupling these receptors to PLA_2 activity (Fig. 2.10).[27,28]

GROUP I AND II PHOSPHOLIPASES A_2

Low molecular-weight, group I PLA_2 comprises the well-studied pancreatic enzyme first isolated in 1932 by V. Gronchi in Italy,[29] which is secreted into the intestine and whose major physiological function is to digest emulsified phospholipids. In addition to the pancreas, group I PLA_2 may also be expressed in other organs, such as the lungs, where a digestive role is unlikely. Evidence that extrapancreatic group I PLA_2 may also participate in cellular signaling comes from the intriguing discovery, by Hitoshi

Arita and his coworkers in Japan, that administration of this enzyme produces an array of cellular responses—ranging from vascular smooth muscle contraction to fibroblast proliferation—by binding with nanomolar affinity to a specific cell membrane receptor (which, as we shall see later on in this section, also binds to group II PLA_2). These multiple effects appear, however, to be independent of phospholipid hydrolysis, and are therefore outside the scope of this book.[30-33]

Instead, we shall now turn to group II PLA_2, a low molecular weight PLA_2 for which evidence indicates an involvement in stimulus-dependent arachidonate mobilization. The structural and mechanistic properties of this PLA_2 have been exhaustively studied. It is composed of 124 amino acid residues, it has a molecular mass of

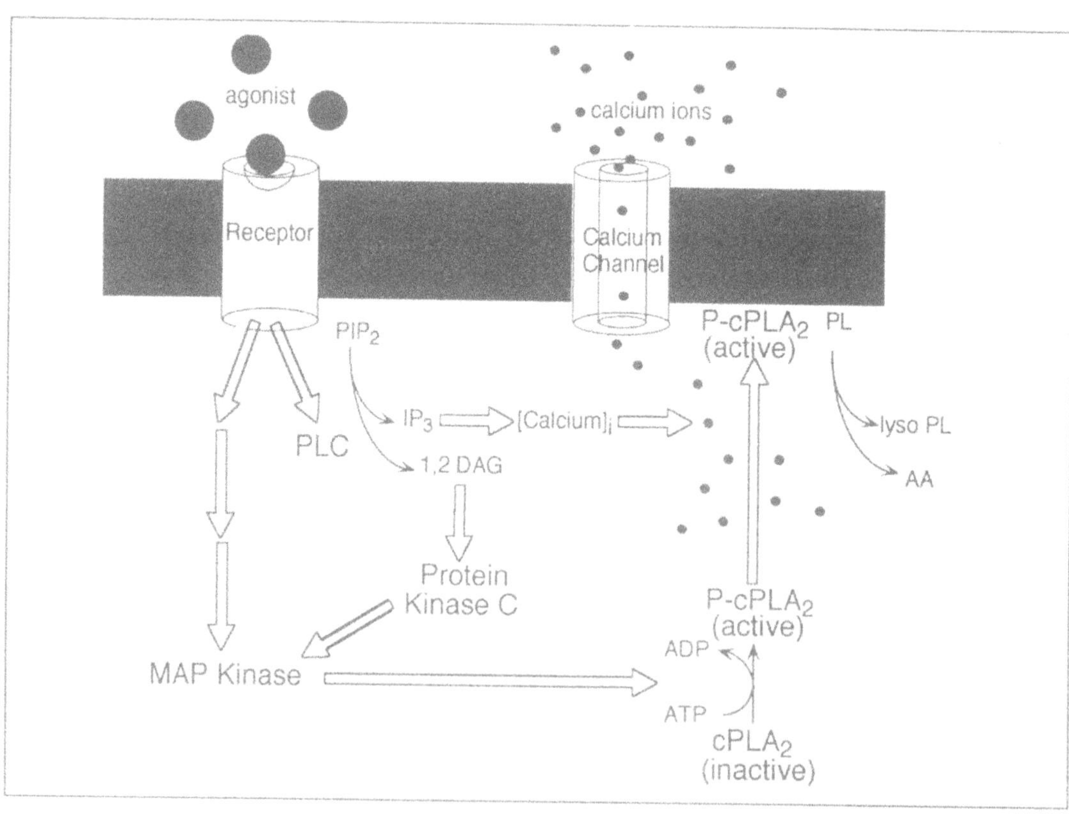

Fig. 2.5. Schematic diagram of the regulation of $cPLA_2$ activity by receptor stimulation and intracellular calcium rises. Calcium binding, phosphorylation and subsequent translocation of $cPLA_2$ are thought to be essential steps in the activation mechanism of this phospholipase, which selectively cleaves arachidonate (AA)-containing phospholipids (PL). $cPLA_2$ phosphorylation may be brought about by the receptor-dependent stimulation of MAP-kinase, phospholipase C (PLC) or calcium channel activities.

about 14 kDa and it is encoded in humans by a single gene. One of its salient structural characteristics is the presence of seven disulfide bridges, which account for its sensitivity to reducing agents, such as dithiotreitol, and for its resistance to heat or acid treatment. Group II PLA_2 is active at neutral to alkaline pH values (with a pH optimum between 8 and 10), requires millimolar calcium as a cofactor, and it shows a marked preference for phospholipid substrate presented in the form of bacterial membranes (for instance, radioactively labeled *Escherichia coli* membranes).[34-39]

X-ray diffraction studies of crystallized group II PLA_2 from various sources, including mammalian tissues, have led to the proposal of a two-step mechanism of catalysis. First, the enzyme binds to the water-lipid interface; indeed, the equilibrium between soluble and membrane-associated enzyme is thought to be the cause of its unusual kinetic behavior. Second, a single molecule of phospholipid substrate positions itself within the enzyme's active site. The calcium ion bound to the enzyme (and essential for its activity) may be instrumental in both aligning the phospholipid with the enzyme's hydrophobic pocket (by binding to the *sn*-3 phosphate group) and in activating the phospholipid substrate (by binding to the carbonyl oxygen at the *sn*-2 fatty acyl group).[14,15]

Group II PLA_2 is found, among other tissues, in blood platelets, spleen, liver and brain, where it is present both in soluble form and in association with membranes.[40] The soluble form is released into the extracellular fluid in a stimulus-dependent manner (Fig. 2.6). Inflammatory stimuli are particularly effective in eliciting secretion of group II PLA_2 and, during inflammatory events, this enzyme can be found outside cells at very high concentrations (e.g., in synovial fluid, ascitic fluid and serum). Also, platelets challenged with the procoagulating protein, thrombin, secrete group II PLA_2.[41,42] In addition to this involvement in pathological conditions, and pointing to an evolutionarily ancient role in intercellular communication, is the fact that PLA_2 secretion (and the arachidonate mobilization that accompanies it) are key-steps in the establishment of cell to cell contacts in the sponge, *Geodia cydonium*.[43] Whether analogous mechanisms occur and play similar roles in vertebrates, is an exciting possibility that remains to be explored.

Association of group II PLA_2 with membranes may entail, at least in the case of the released enzyme, binding to a high affinity receptor—a transmembrane protein of ≈180 kDa, expressed predominantly in embryonic tissues and able to recognize both group II PLA_2 (with a k_D of ≈0.8 nM) and group I PLA_2 (k_D ≈10 nM).[44,45] The receptor-bound enzyme may be catalytically active, and some (but not all) of its effects may be mediated through arachidonate mobilization and eicosanoid formation (Fig. 2.6).

In some cases, association of group II PLA_2 with cells appears to be unrelated to their non-covalent binding with membrane proteins or polysaccharides. This cell-associated form of group II PLA_2 may also contribute to receptor-stimulated arachidonate mobilization

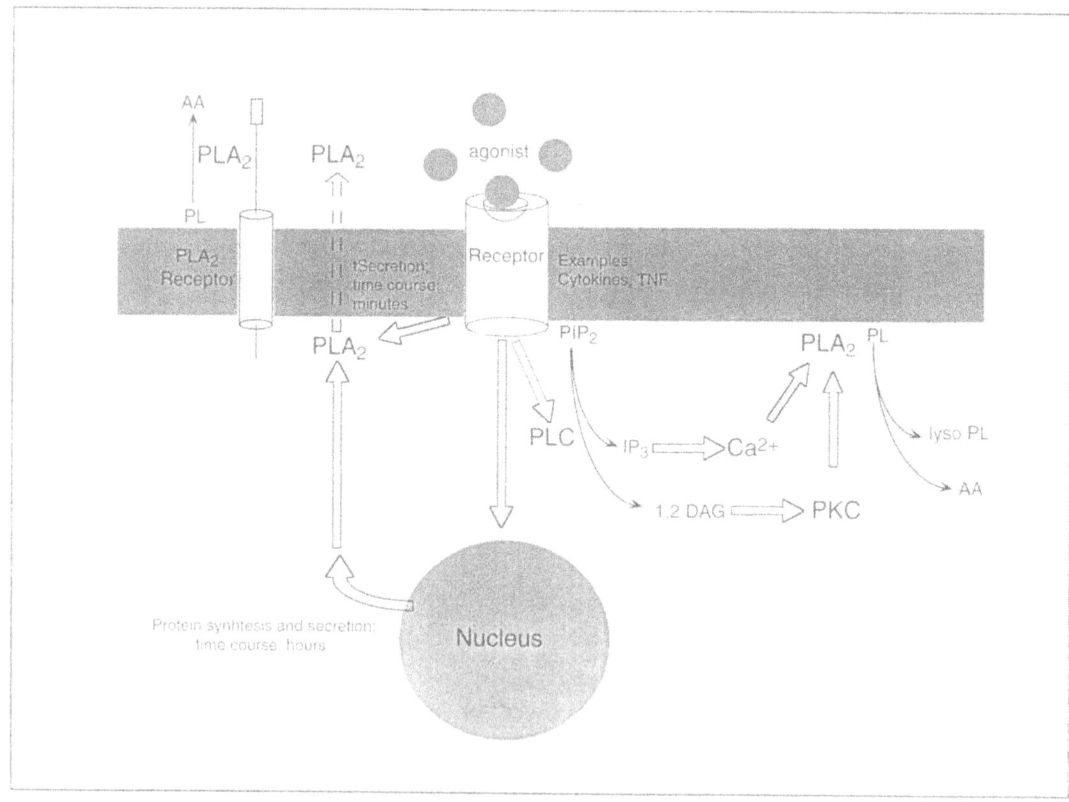

Fig. 2.6. Possible mechanisms for the receptor-dependent activation of type II PLA_2 in mammalian cells. Short-term mechanisms include the stimulation of intracellular calcium rises (shown on the right) and the stimulation of type II PLA_2 secretion (shown on the left). Long-term enhancement of type II PLA_2 activity may be achieved through stimulation of PLA_2 synthesis. Extracellular PLA_2 is thought to act by binding to a membrane-spanning high-affinity receptor.

(Fig. 2.6): in transfected cells that overexpress group II PLA_2, the activation of transducing G proteins with fluoroaluminate (which mimics GTP in stimulating G-protein activity) or of protein kinase C with phorbol esters results in a greater arachidonate release than in control, mock-transfected cells.[46] Other cases in which group II PLA_2 does not seem to participate in stimulus-dependent arachidonate mobilization have been documented.[47]

CALCIUM-INDEPENDENT PHOSPHOLIPASE A_2

PLA_2s belonging to groups I to IV are all calcium-dependent enzymes. Calcium-independent PLA_2s also exist, however, and may play important roles in arachidonate turnover, though lack of adequate information on their primary structure prevents their assignment to any of the groups listed in Table 2.3.

Calcium-independent PLA_2s have been purified from canine heart tissue, by the laboratory of Richard W. Gross, at the Washington University in St. Louis,[48] and from the macrophage-like cell line P388D1, by the laboratory of Edward A. Dennis, at the University of California in San Diego.[49] Beside calcium-independence, these two enzymes share several noteworthy features: they both are cytosolic, activated by ATP in a non phosphorylation-dependent manner, and form high molecular mass complexes of ≈400 kDa. They differ in kinetic properties as well as in substrate selectivity: the heart enzyme shows a marked preference for plasmalogens containing arachidonic acid at the *sn*-2 position, while the P388D1 enzyme is much less specific in its substrate requirements.[48,49]

In general, calcium-independent PLA_2s coexist in cells with other, calcium-dependent PLA_2 activities, raising the question of their respective roles in cell lipid metabolism and signaling. One possibility is that, while calcium-dependent PLA_2s may be crucial for arachidonate mobilization, their calcium-independent counterparts may provide the lysophospholipid substrate necessary for the reacylation of arachidonic acid into phospholipids.

This idea is depicted schematically in Figure 2.7. Incorporation of arachidonic acid into phospholipids depends on arachidonoyl-CoA:lysophospholipid transferase activity. Because arachidonate is converted very rapidly to arachidonoyl-CoA within cells, this reaction is unlikely to be rate-limiting. But what about the lysophospholipid? We know that the concentrations of this intermediate increase when calcium-dependent PLA_2s are activated. Yet, arachidonic acid is quickly esterified even in resting cells, when calcium-dependent PLA_2 activities are negligible. Edward A. Dennis and his coworkers have provided a likely solution to this conundrum by demonstrating the existence of a calcium-independent PLA_2 activity whose major role may be to supply lysophospholipids for the biosynthesis of arachidonoyl-phospholipids (Fig. 2.7).[50]

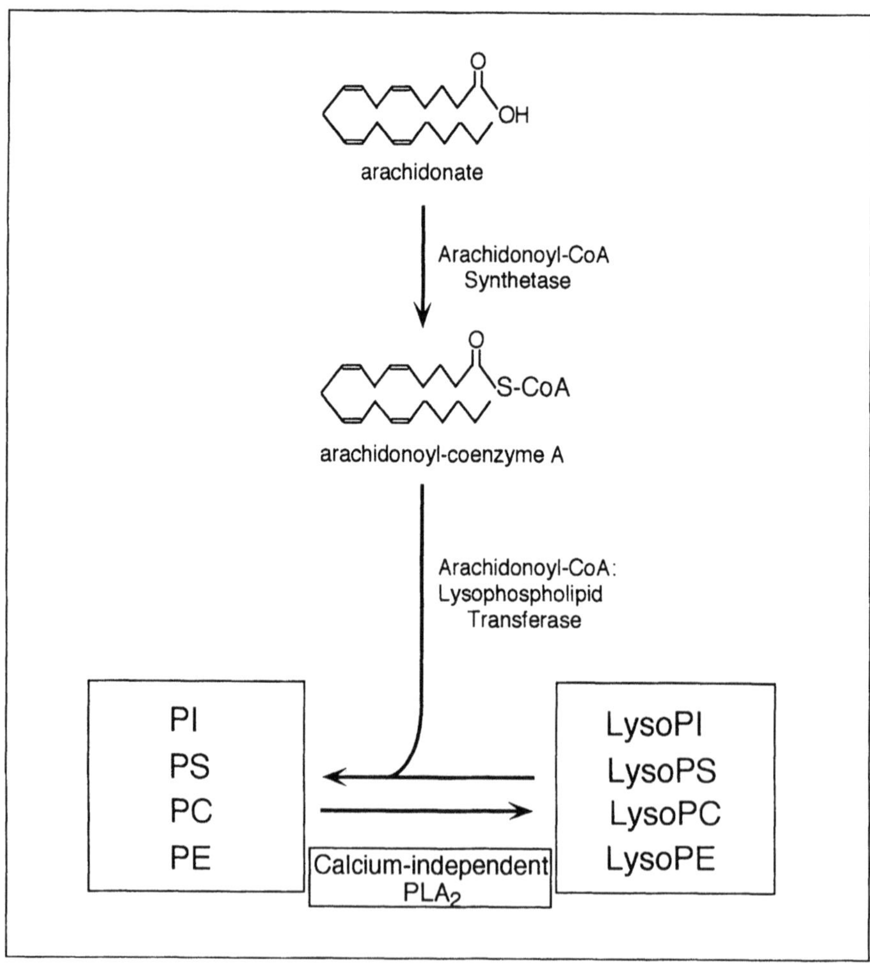

Fig. 2.7. Possible participation of cytosolic, calcium-independent PLA_2 in the formation of arachidonoyl-containing phospholipids. By producing lysophospholipids, calcium-independent PLA_2 may set the stage for the activity of arachidonoyl-CoA:lysophospholipid transferase. These reactions would result in enriching the arachidonate content of phospholipids at the sn-2 position.

As one would expect, for all its elegance, this model does not exhaust the potential functions of calcium-independent PLA_2s. Evidence tallies also with a direct participation of this enzyme in arachidonate mobilization: in pancreatic β cells, for example, an ATP-activated, "heart-type" isoform of calcium-independent PLA_2 may mediate mobilization of arachidonate and subsequent formation of 12-lipoxygenase metabolites, which are thought to act as intracellular mediators in glucose-induced insulin secretion.[51]

PROTEINS THAT MODULATE PHOSPHOLIPASE A_2 ACTIVITY

PLA_2 activity may be stimulated by a polypeptide called PLA_2-activating protein, or PLAP. PLAP, purified and cloned by Mike A. Clark and collaborators (at the SmithKline and French Laboratories), is a ≈28 kDa protein that cross-reacts with antibodies raised against the bee venom peptide, mellitin.[52,53] Like mellitin, with which it shares some sequence homology, PLAP activates mammalian group II PLA_2 activity whereas it has no effect on group I and group IV PLA_2s.[54] The expression of PLAP is increased by some inflammatory mediators, such as tumor necrosis factor (TNF) and leukotriene D_4.[55]

Several protein inhibitors of PLA_2 activity have also been described. A great deal of work has focused on a class of proteins called annexins (also referred to in the older literature as lipocortins), which bind to phospholipids in a calcium-dependent fashion. The annexins are found on the cytoskeletal network that underlies the plasma membrane and are thought to inhibit PLA_2 activity by sequestering its phospholipid substrate. Although the identity of the PLA_2 isoform modulated by the annexins has not been determined, indirect evidence suggests that this may belong to group II enzymes (for review, see ref. 16).

THE PHOSPHOLIPASE C/ACYLGLYCEROL-LIPASE PATHWAY

Activation of the various PLA_2 isoforms described in the preceding sections provides a single-step pathway of arachidonate mobilization. In addition to this direct mechanism, arachidonate mobilization may be initiated by the stimulation of PLCs, which catalyze the hydrolysis of phospholipids (mainly phosphoinositides and phosphatidylcholine) at the phosphate ester bond, producing 1,2 diacylglycerol and a phosphorylated polar head compound (Fig. 2.4 and 2.8; this enzymatic activity is also referred to as "phosphodiesterase"[56]).

1,2 Diacylglycerol is a familiar second messenger, best known for its ability to activate several members of the PKC family. Like most lipid messengers, though, 1,2 diacylglycerol is a Janus-faced molecule: second messenger on the one hand, a precursor for other biologically active lipids—including arachidonate—on the other.

The metabolic fate of 1,2 diacylglycerol is shown in Figure 2.8. The diglyceride is hydrolyzed by a stereoselective diacylglycerol lipase activity which releases the fatty acid at the *sn*-1 position (usually a saturated fatty acid, such as stearic acid), and forms 2-acylglycerol (usually containing a polyunsaturated fatty acyl group, such as arachidonic acid). 2-Arachidonoylglycerol (in itself a bioactive substance, see chapter 6) is cleaved by a monoacylglycerol lipase activity, forming free arachidonate and glycerol. This series of reactions, discovered in blood platelets by Robert L. Bell and Philip W. Majerus,[57] has been proposed to mediate arachidonate mobilization in various tissues and cells; for instance, the mobilization induced by bradykinin in primary cultures of sensory neurons.[58]

In those experiments, neurons obtained from the spinal cord of embryonic rats were labeled by incubation with various radioactive lipid precursors, and exposed to bradykinin. Application of this peptide raised the levels of unesterified arachidonate, but had no effect on lysophospholipids, arguing against an involvement of PLA_2. By contrast, appearance of the free fatty acid was preceded by a transient increase in 1,2 diacylglycerol content which took place within a few seconds of exposure to bradykinin. Because of its rapid time-course, this rise in 1,2 diacylglycerol may derive from the direct stimulation of a PLC activity.

It should be added, however, that the appearance of 1,2 diacylglycerol is, in general, insufficient to conclusively infer an involvement of the PLC/acylglycerol pathway: the most resilient difficulty

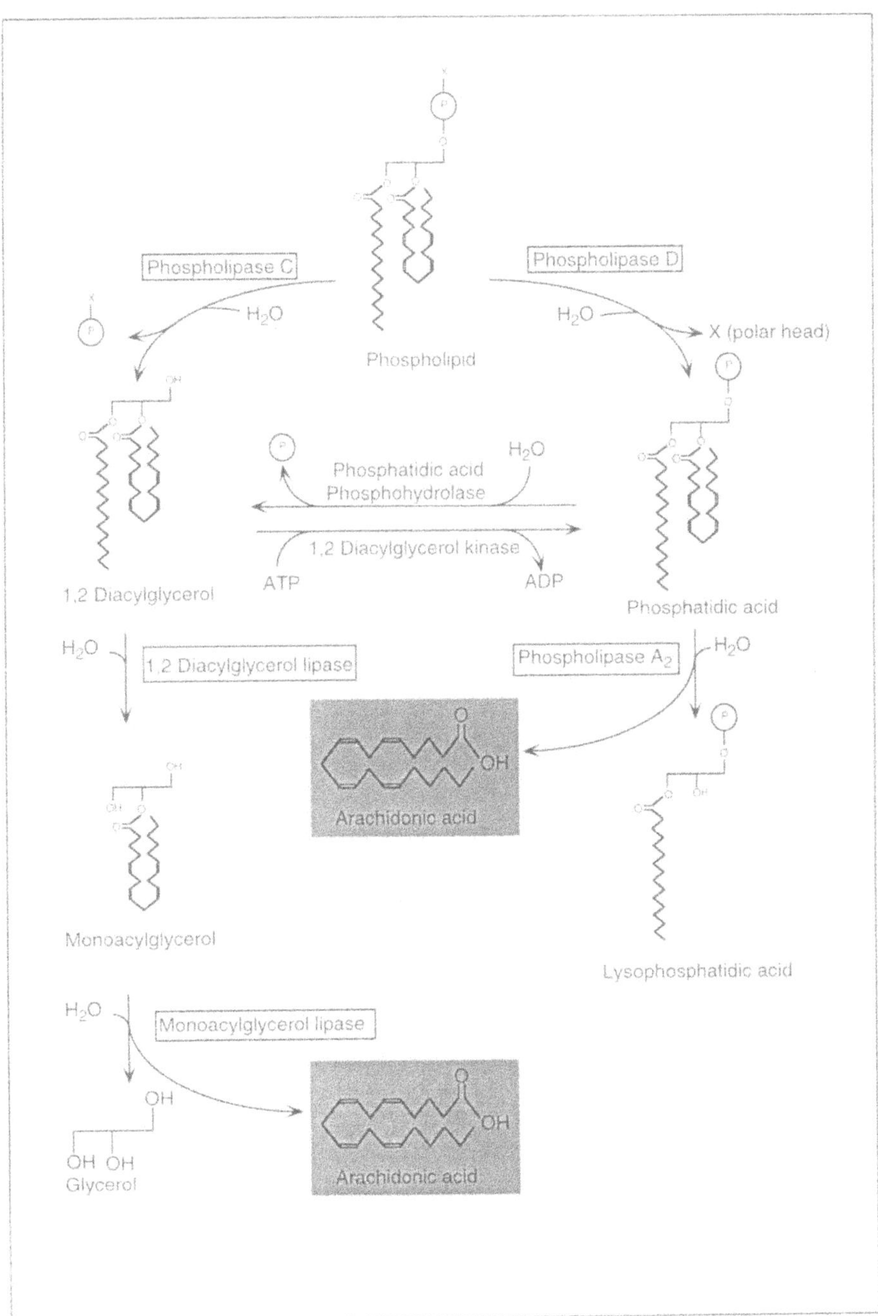

Fig. 2.8. Arachidonate mobilization through the phospholipase C and phospholipase D pathways.

in defining whether this pathway mediates arachidonate mobilization is in the rapid metabolic interconversions that link intermediates in the PLC cascade with those of another important phosphodiesterase, PLD (Fig. 2.8).

PHOSPHOLIPASE D

Hydrolysis of phospholipids by PLDs—a family of enzymes of which none has yet been purified to homogeneity (but this is probably going to change in the near future)—produces phosphatidic acid and a non-phosphorylated polar head compound, such as choline or ethanolamine (Fig. 2.4 and 2.8).[59,60] Needless to say, phosphatidic acid is biologically active (among other things, it stimulates protein phosphorylation), but we shall focus here only on its role as arachidonic acid precursor. Two enzymatic pathways are involved (Fig. 2.8).

A phosphohydrolase activity cleaves the phosphate ester bond of phosphatidic acid forming 1,2 diacylglycerol, which may then enter its usual metabolic route; the reaction goes both ways, and phosphatidic acid may be synthesized from 1,2 diacylglycerol by the action of a lipid kinase (Fig. 2.8). As mentioned, this two-way shunt plays havoc with the attempts to determine unequivocally whether 1,2 diacylglycerol and phosphatidic acid derive from the activation of PLC or PLD.[59]

Phosphatidic acid may also be hydrolyzed by a PLA_2 (Fig. 2.8). One such enzyme, recently purified from rat brain, has an apparent molecular mass of ≈58 kDa (as determined by SDS-PAGE), and shows little or no catalytic activity on phosphatidylcholine, phosphatidylethanolamine and diacylglycerol.[61] It is important to point out that the two products of PLA_2 activity on phosphatidic acid, arachidonic acid and lysophosphatidic acid, are both thought to act as signaling molecules: lysophosphatidic acid binds with high affinity to a G protein-coupled membrane receptor and may serve therefore a transcellular messenger function.[62]

Under what physiological conditions may PLC and PLD participate in stimulus-dependent 1,2 diacylglycerol formation, and hence, create the conditions for arachidonate mobilization? From what I have just said, a complete and satisfactory answer to this question is still lacking, but the current general consensus may be summarized as follows. When membrane receptors are activated, cells produce 1,2 diacylglycerol in a biphasic manner. The first phase, which is transient (it peaks usually within 30 seconds of stimulation) and quantitatively small, coincides with phosphoinositide hydrolysis by PLC. The second phase, which may take place independently of the first, is long-lasting (a peak is reached between 2 and 15 minutes of stimulation), results mainly from the hydrolysis of phosphatidylcholine, and is mediated in some cells by PLC (e.g., MDCK-D1 kidney cells or 3T3-L1 fibroblasts) and in others by PLD (e.g., neutrophils). This somewhat Solomonian conclusion may not be intellectually satisfying, but it is realistic: the existence of a clear-cut division of labor between PLC and PLD in lipid signaling appears at present unlikely.

INHIBITORS OF ARACHIDONATE MOBILIZATION

The researcher who wishes to tackle the role of specific phospholipases in arachidonate mobilization can turn to a drug arsenal that is much more richly supplied today than just a few years ago. Potent, selective and membrane-permeant inhibitors for several groups of PLA_2s are now available; some of these agents are listed in Table 2.4. New drugs for group II PLA_2 have largely replaced old and non-selective favorites such as mepacrine (which acts by interfering with the phospholipid substrate) or p-bromophenacyl bromide (a histidine-alkylating agent). Promising inhibitors of group IV PLA_2 as well as of cytosolic, calcium-independent PLA_2 have also been described (Table 2.4).[63-65] In addition to these enzyme inhibitors, anti-sense oligodeoxynucleotides targeted against specific PLA_2 sequences have been used with success in a number of experiments in vitro to prevent the expression of select PLA_2 isoforms.[66-69] The potential therapeutic applications of these two different class of inhibitors are vast—combating inflammation being a classical example though probably not the only one—and are currently being explored.

RECEPTOR-DEPENDENT CONTROL OF ARACHIDONATE MOBILIZATION

I have briefly described the structural, catalytic and pharmacological properties of various enzymes implicated in arachidonic acid

Table 2.4. Select inhibitors of arachidonic acid mobilization

Enzyme	Inhibitor	IC_{50}	Comments
Group I PLA_2	Oleoyl oxyethylphosphorylcholine	6 μM	Cell impermeable, reversible;
Group II PLA_2	p-Bromo-phenacyl-bromide	43 μM	Cell permeable, irreversible;
	manoalide	0.017 μM	Cell permeable, irreversible, inhibits also Type I PLA_2;
	scalaradial	0.017 μM	Cell permeable, irreversible;
	[4-(3,5-didodecanoyl-2,4,6-trihydroxyphenyl) -7-hydroxy-2(hydroxyphenyl)chroman [compound YM-26734]	0.085 μM	Selective for Group II PLA_2 over Group I and IV , cell permeable;
Group IV PLA_2 [$cPLA_2$]	arachidonoyltrifluoromethylketone	NA	Selective for Group IV PLA_2 , cell permeable;
calcium-independent PLA_2	(E)-6-(bromomethylene0tetrahydro-3-(1-naphathalenyl)-2H-pyran-2-one	0.060 μM	Selective for calcium-independent PLA_2, cell permeable;
PLC, pl-specific	1-O-octadecyl-2-O-methyl-sn-glycero-3-phosphorylcholine	0.4-10 μM	
Diacylglycerol lipase	1,6-bis-(cyclohexyloximinocarbonylamino) -hexane [compound RHC-8267]	4 μM	

biosynthesis and turnover. In the following sections, we will be concerned with another essential question: how do living cells use this array of enzymes to control free arachidonate levels, when an appropriate stimulus excites them?

We have seen that a series of 'first messenger' signaling molecules produce arachidonate mobilization by activating membrane receptors. Too many such effects have been described over the past 20 years to allow me to compile a review of this subject area that is both exhaustive and useful. I will rather illustrate with some examples what I consider to be the three major modalities of receptor-mediated control of arachidonate mobilization: stimulation, inhibition and facilitation. Where possible, I will also indicate the specific transmembrane signaling pathway and enzyme route affected.

RECEPTORS THAT STIMULATE ARACHIDONATE MOBILIZATION

Animal cells have evolved two sorts of mechanisms to increase arachidonate mobilization in response to external stimuli: short-term mechanisms, which evolve over a time-scale of seconds to minutes and do not depend on the synthesis of new proteins, and long-term mechanisms, which evolve over several hours and do depend on protein synthesis.

I will provide two examples of short-term arachidonate mobilization which involve distinct transduction systems and different PLA_2 activities: the activation of group IV PLA_2 by purinergic receptors in Chinese hamster ovary (CHO) fibroblasts, and the activation of group II PLA_2 by platelet-activating factor (PAF) or bacterial lipopolysaccharide in $P388D_1$ cells, a macrophage-derived cell line.

In CHO cells, the stimulation of constitutive P_2-type purinergic receptors by extracellular ATP produces two prominent responses: activation of phosphoinositide-specific PLC, which leads to inositoltrisphosphate-induced elevations of intracellular calcium levels, and formation of free arachidonate from membrane phospholipids. Evidence from both pharmacological and genetic experiments, carried out in the laboratory of Gary L. Johnson (at the University of Colorado Medical School), indicates that these intracellular signaling pathways are by and large independent.[70] Pharmacologically, the two pathways can be inhibited differentially with pertussis toxin, a bacterial toxin that covalently modifies and inhibits the α_i and α_o subunits of heterotrimeric G proteins: arachidonate mobilization is very sensitive to the toxin, whereas PLC stimulation is not. Genetically, CHO cells can be modified to express a G protein mutant that is unable to couple P_2 receptors to arachidonate mobilization, but leaves unhindered the receptor-dependent activation of PLC.

From these results, one may reasonably conclude that P_2-type purinergic receptors are coupled directly to arachidonate mobilization (i.e., that mobilization is not secondary to activation of PLC, although it depends on the calcium rise produced by inositoltrisphosphate) and that a transducing G protein is an integral component of this coupling mechanism (Fig. 2.9). What phospholipase activity is involved in this reaction? To address this question, Lih-Ling Lin and her coworkers (at the Genetics Institute in Cambridge, Massachusetts) have prepared clones of CHO cells that stably overexpress either group IV or group II PLA_2s, and determined the effect of this genetic manipulation on receptor-dependent arachidonate mobilization.[71] In cells overexpressing group II PLA_2, the effect of purinergic receptor stimulation was similar to that of nontransfected cells, whereas in cells overexpressing group IV PLA_2 this effect was strongly enhanced. Further support for an involvement of group IV PLA_2 came from experiments in which expression of group IV PLA_2 was inhibited by incubating CHO cells with synthetic oligonucleotides: cells treated with anti-sense-oriented oligonucleotides showed little or no P_2 receptor-dependent arachidonate mobilization, whereas mobilization was normal in control cells treated with either sense-oriented or mis-sense ('scrambled') oligonucleotides.[69]

Let us turn now to our second example of short-term arachidonate mobilization: the receptor-dependent activation of group II PLA_2 in $P388D_1$ cells. Stimulation of these cells with lipopolysaccharide or PAF mobilizes arachidonate in two phases: a transient phase (maximal at ≈1.5 minutes and rapidly declining afterward) in which arachidonate accumulates within the cells, and a longer-lasting phase (maximal at ≈10 minutes and persistent for up to 30 minutes in the presence of receptor agonist) in which the fatty acid accumulates in the incubation medium. Jesùs Balsinde, Edward A. Dennis and collaborators were able to show that the long-lasting phase of arachidonate mobilization can be prevented by treating P388D_1 cells with anti-sense oligonucleotides to group II PLA_2.[72] These oligonucleotides have no effect on the early transient phase, which the authors suppose therefore to be mediated by cytosolic group IV PLA_2.

How may activation take place under those conditions? We don't know yet, but a possible scenario (composed from results

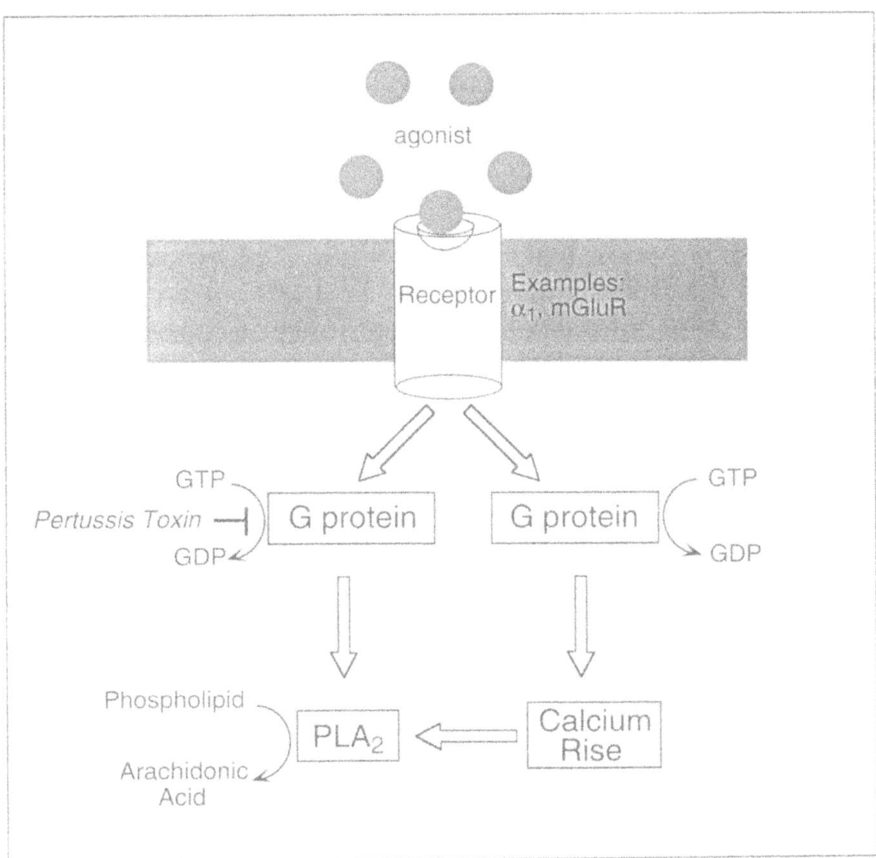

Fig. 2.9. Regulation of PLA_2 activity by heterotrimetric G proteins: receptor-dependent activation. This hypothetical scheme combines results from several lines of experimentation, carried out mainly on heterologous expression systems.

obtained in different systems) is illustrated in Figure 2.6. Two distinct mechanisms are hypothesized: one, in which cell-associated group II PLA_2 becomes active in response to a receptor-operated transmembrane signal (this is not known yet, but it may entail calcium rises and/or PKC activation); another, in which receptor occupation results in the secretion of group II PLA_2. Once outside the cell, the enzyme may bind to its own high-affinity membrane receptor (either on the cell of origin or on an adjacent cell) and find the high calcium concentration necessary to support its activity.

Figure 2.6 outlines also a third possibility, i.e. that receptor occupation may send a signal to the nucleus for the synthesis of new PLA_2. An example of this long-term regulatory mechanism is provided by two inflammatory cytokines (interleukin 1_{β} and tumor-necrosis factor α) which, in rat mesangial cells, cause both secretion of preformed group II PLA_2 and expression of new PLA_2.[73] Underscoring the pleiotropic nature of these responses are results indicating that tumor-necrosis factor α may also induce the synthesis of group IV PLA_2[74] and of PLAP, the PLA_2-activating protein.[55]

Growth factors such as PDGF and EGF also have dual effects, short-term and long-term, on group IV PLA_2. Figure 2.10 illustrates how these tyrosine kinase receptors may be linked to a rapid increase in PLA_2 activity, via the MAP kinase cascade, and to a more sustained increase, via gene expression and synthesis of new PLA_2 protein.[27,75]

SOME RECEPTORS MAY INHIBIT ARACHIDONATE MOBILIZATION

By analogy with other transmembrane signaling systems—the cyclic AMP pathway for instance—one would expect the arachidonic acid cascade to be regulated both positively, as we have summarized above, and negatively. While evidence for an inhibitory control has been obtained, it is admittedly still very incomplete. In fact, to the best of my knowledge, only two examples of such control have been reported.

Non-hydrolyzable GTP analogs (e.g., GTP-γ-S) have played a very important role in determining the role of G proteins in transmembrane transduction. As a rule, their ability to produce a certain response is taken as good evidence for the presence of a G protein-mediated coupling mechanism. By using GTP analogs, Carol Jelsema and Julius Axelrod (at the National Institutes of Health in Bethesda) have provided evidence for an inhibitory control by G proteins over PLA_2 activity in the retina.[76,77] While studying signaling events in retinal photoreceptors, they observed that flashing light on dark-adapted rod outer segments (ROS) enhanced PLA_2 activity (unfortunately not characterized in those studies). When the ROS were exposed, however, to light after incubation with GTP-γ-S, this increase was significantly smaller. They concluded

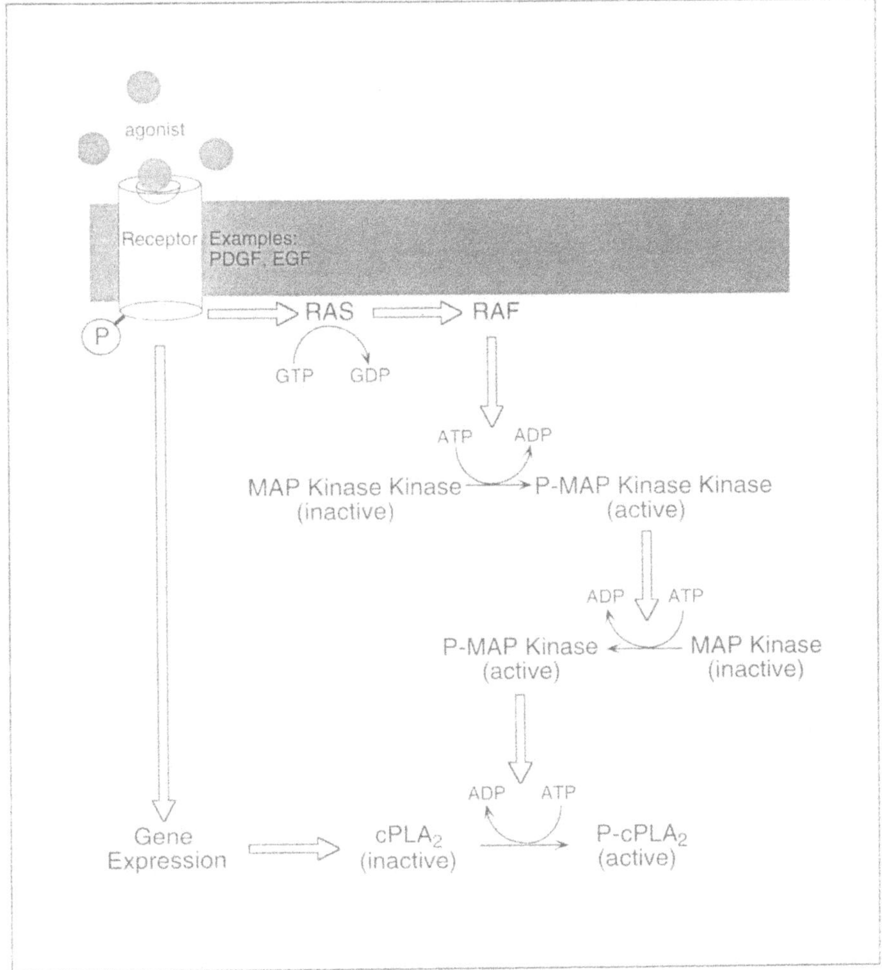

Fig. 2.10. Hypothetical regulation of PLA_2 activity by protein tyrosine kinases.

that an unidentified G protein, which could be activated by the GTP analog, exerted an inhibitory action on the activity of retinal PLA_2 when this enzyme was stimulated by light (for contrasting results, see ref. 83).

One experiment suggests that transmembrane receptors may also link to inhibition of arachidonate mobilization.[78] CHO cells were transfected with a plasmid vector directing expression of the H_2-type histamine receptor, which is known to be positively linked to adenylyl cyclase, via a G_s transducing protein. CHO cells were no exception to this rule, and the transfected receptor turned out to be very effective in increasing cyclic AMP levels when stimulated with an H_2 agonist. Quite unexpectedly, though, H_2-receptor occupation was also found to reduce the mobilization of arachidonate produced by either purinergic receptor stimulation or application of a calcium ionophore. The mechanism of this response was not determined, but two obvious possibilities were ruled out. First, inhibition of arachidonate formation was not secondary to decreased calcium rises because H_2-receptor stimulation had no effect on either basal or stimulated calcium levels. Second, inhibition was independent of the stimulating effect of H_2 receptors on cyclic AMP levels because membrane-permeant cyclic AMP analogs had no such effect. G-protein mediation was not determined in these studies.

It appears from these results that certain transmembrane receptors may have the potential to be negatively coupled to the arachidonate cascade, at least under the artificial conditions of an heterologous expression system. Whether analogous responses may occur with receptors constitutively expressed in cells and tissues, remains a matter of conjecture.

A THIRD GROUP OF RECEPTORS FACILITATE ARACHIDONATE MOBILIZATION, BUT DO NOT STIMULATE IT DIRECTLY

Several neurotransmitter receptors share the ability to decrease adenylate cyclase activity in cells, through the intermediate of an α_i transducing G protein. When transfected in competent mammalian cells (e.g., CHO cells), receptors of this group produce, in addition to this classical effect on cyclic AMP accumulation, what may be described as a "silent" facilitation of arachidonate mobilization. As shown schematically in Figure 2.12, activation of such receptors has no effect per se on arachidonate mobilization, but it greatly potentiates the mobilization evoked by a second agent—for example, by stimulating another receptor or by applying a calcium ionophore (see also Fig. 2.13 A). Neurotransmitter receptors belonging to this group include dopamine D_2, serotonin 5-HT_1, adrenergic α_2, muscarinic m_2 and m_4, and adenosine A_1 receptors (reviewed in ref. 79).

In many cases, this facilitatory effect on arachidonate mobilization requires, like adenylate cyclase inhibition, the participation

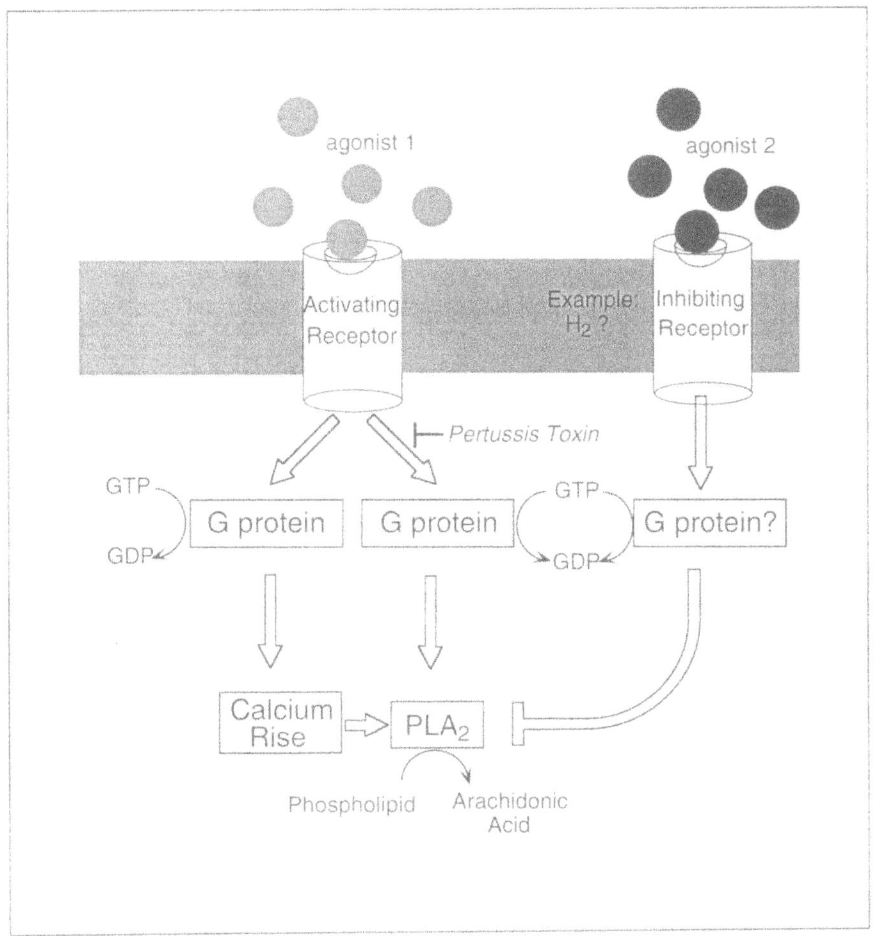

Fig. 2.11. Regulation of PLA_2 activity by G proteins: receptor-dependent inhibition. Hypothetical scheme drawn from results reported in ref. 78.

of an $G\alpha_i/G\alpha_o$ protein. This conclusion is supported by experiments carried out on CHO cells transfected with a cDNA encoding the D_2 dopamine receptor: in these cells pertussis toxin prevented the response, while the stable GTP analog, GTP-γ-S, mimicked it (after permeabilization of the cells).[80,81] In the same expression system, arachidonate mobilization is mediated, as we have seen, by activation of group IV PLA_2. In agreement with this possibility, anti-sense oligonucleotides directed against group IV PLA_2 prevent the D_2 receptor-dependent facilitation of arachidonate release, whereas they have no effect on several other intracellular signaling pathways.

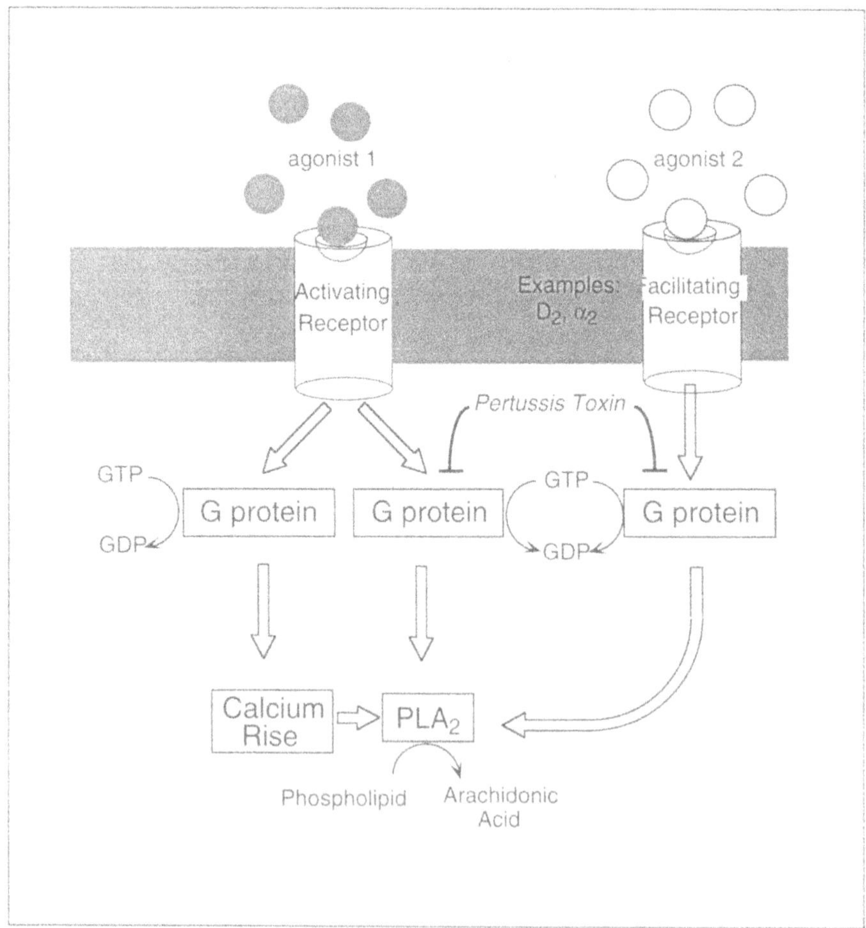

Fig. 2.12. Regulation of PLA_2 activity by G proteins: receptor-dependent facilitation. Discovered in heterologous expression systems, this mechanism of PLA_2 activation has been recently demonstrated also in primary cultures of rat brain neurons (see ref. 79 for review).

SECOND MESSENGER INTERACTIONS ON ARACHIDONATE MOBILIZATION

The arachidonic acid cascade takes part in a large number of functional interactions with other intracellular second messenger systems. As one would expect from their universal regulatory functions in the physiology of mammalian cells, protein phosphorylation and dephosphorylation play a major part in modulating arachidonate mobilization. Two multifunctional (and ubiquitous) protein kinases, PKC and PKA, have been studied extensively in this regard.

Activation of PKC causes in most cells an increased stimulus-dependent mobilization of arachidonate. As we have seen, the mechanism of this enhancement is not entirely understood, but it likely involves a cascade of biochemical events leading to the phosphorylation and activation of cytosolic group IV PLA_2.[25,26] The possible significance of this modulatory effect—a 'priming' of cells in view of an increased response to upcoming stimuli—has been pointed out, yet some of its offshoots may also prove to be of some interest.

PKC-dependent upregulation of PLA_2 activity may participate, for example, in 'remodeling' the spectrum of intracellular signals driven by the stimulation of dopamine D_2 receptors. These receptors, which exert a modulatory control on the excitability of mammalian neurons and are also present in peripheral tissues, are linked to the parallel activation of several G proteins and to the formation of multiple intracellular signals. These include inhibition of adenylate cyclase activity and facilitation of evoked arachidonate mobilization (see above). Evidence from CHO cells transfected with D_2 receptors indicates that PKC activity may control effector coupling at this receptor by switching it from a prevalent inhibition of adenylate cyclase to a prevalent facilitation of arachidonate mobilization.[81] By doing so, PKC may cause the recruitment of a distinct spectrum of biologically active mediators in cells in which PKC activity has been elevated. Through such a mechanism, the physiological responses to dopamine could be regulated dynamically within a cell by stimuli promoting PKC activation (e.g., other neurotransmitters).

Unlike PKC, the effects of PKA activity on arachidonate mobilization can be either stimulatory or inhibitory, and they largely depend on the cell type and on the nature of the stimulus. Moreover, the available data make it difficult to draw any conclusion as to the biochemical mechanism of these effects, except perhaps that they are not likely to entail the direct phosphorylation of a PLA_2.

In one recent study, CHO cells were used to determine whether cyclic AMP-linked receptors may affect the facilitation of arachidonate formation caused by stimulating transfected dopamine D_2 receptors.[80] CHO cells were modified genetically to co-express on their surface two dopamine receptors, the D_1 type, which is positively coupled to adenylate cyclase activity, and the D_2 type, which inhibits the cyclase and facilitates arachidonate mobilization. When both receptors were stimulated with the natural ligand, dopamine, arachidonate mobilization was much greater than when the D_2 receptor alone was stimulated (e.g., in the presence of the selective D_1-receptor blocker, SCH23390, Fig. 2.13). Like all experiments carried out in heterologous expression systems, these experiments have mainly an heuristic value, and need to be interpreted with caution. They do raise, however, an intriguing possibility; namely, that two receptors, traditionally considered as antagonistic for their effects on cyclic AMP levels, may in fact act synergistically on other second messenger pathways. This effect may provide a plausible molecular basis for certain forms of physiological synergism between D_1 and D_2 receptors, and possibly other G protein-linked receptors.[82]

Fig. 2.13. (Opposite page) Dopamine receptor synergism on arachidonate mobilization in transfected Chinese hamster ovary (CHO) fibroblasts. (Modified with permission from Piomelli D et al. Nature 1991; 353:164-167.)

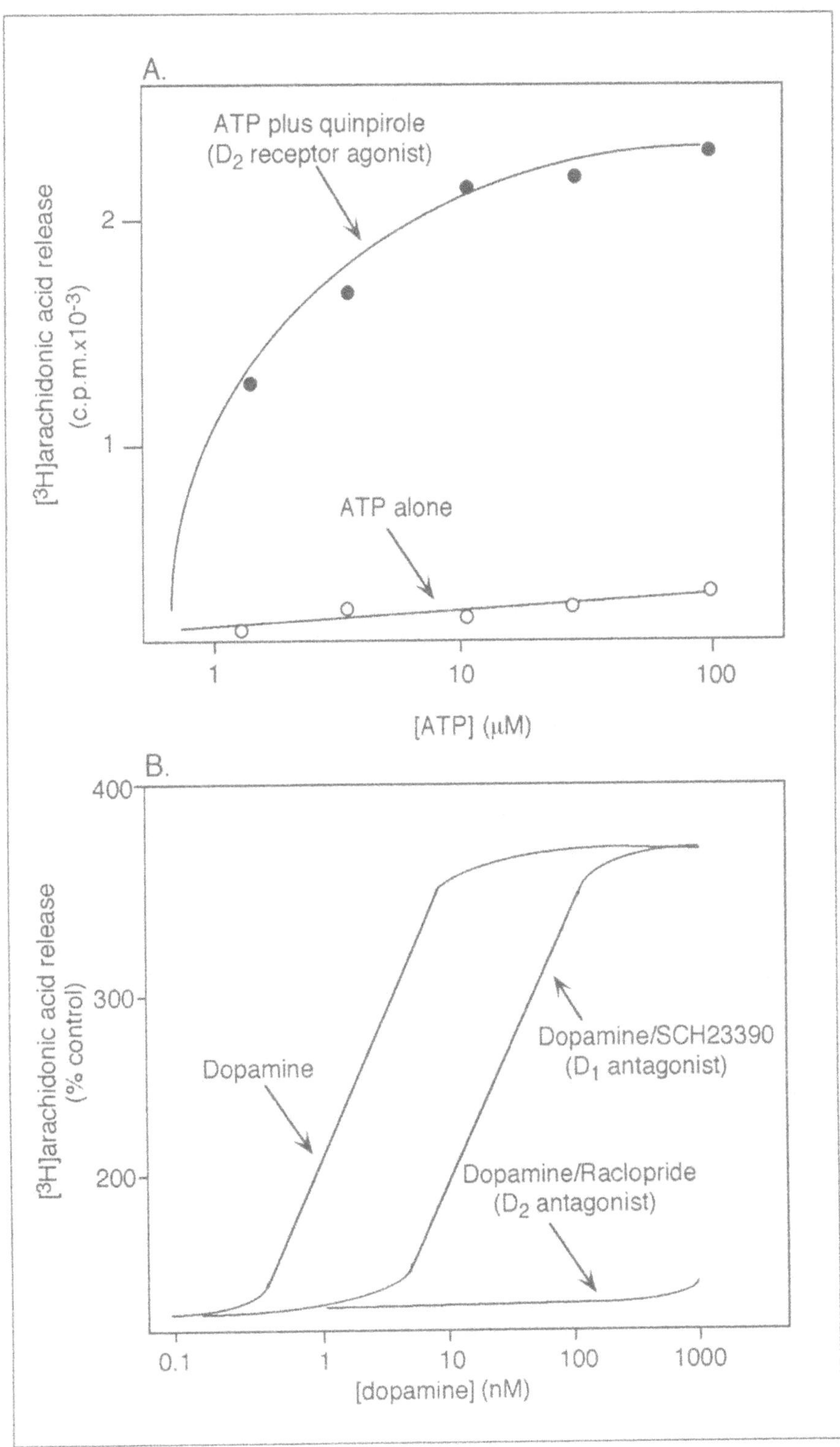
A.
ATP plus quinpirole
(D2 receptor agonist)
ATP alone
[3H]arachidonic acid release
(c.p.m.x10-3)
2
1
1
10
100
[ATP] (μM)
B.
400
300
200
[3H]arachidonic acid release
(% control)
Dopamine
Dopamine/SCH23390
(D1 antagonist)
Dopamine/Raclopride
(D2 antagonist)
0.1
1
10
100
1000
[dopamine] (nM)

REFERENCES

1. Kolesnick R, Golde DW. The sphingomyelin pathway in tumor necrosis factor and interleukin signaling. Cell 1994; 77:325-328.
2. Wakil SJ, Barnes EM. Fatty acid metabolism. Comprehensive Biochem 1971; 18S:57-104.
3. Moore SA, Yoder E, Spector A. Role of the blood-brain barrier in the formation of long-chain ω-3 and ω-6 fatty acids from essential fatty acid precursors. J Neurochem 1990; 55:391-402.
4. Moore SA, Yoder E, Murphy S, Dutton GR, Spector A. Astrocytes, not neurons, produce docosahexaenoic acid (22:6 ω-3) and arachidonic acid (20:4 ω-6). J Neurochem 1991; 56:518-524.
5. Washizaki K, Smith QR, Rapoport SI, Purdon AD. Brain arachidonic acid incorporation and precursor pool specific activity during intravenous infusion of unesterified [^{3}H]arachidonate in the anesthesized rat. J Neurochem 1994; 63:727-736.
6. Spector AA, Hoak JC, Warner ED, Fry GL. Utilization of long chain fatty acids by human platelets. J Clin Invest 1970; 49: 1489-1496.
7. Wilson DB, Prescott SM, Majerus PW. Discovery of an arachidonoyl coenzyme A synthetase in human platelets. J Biol Chem 1982; 257:3510-3515.
8. Laposata M, Reich EL, Majerus PW. Arachidonoyl-coA synthetase. Separation from nonspecific acyl-coA synthetase and distribution in various cells and tissues. J Biol Chem 1985; 260:11016-11020.
9. Irvine R. How is the level of free arachidonate regulated? Biochem J 1982; 204:3-16.
10. Waku, K. Origins and fates of fatty acyl-coA esters. Biochim Biophys Acta 1992; 1124:101-111.
11. Blank ML, Smith ZL, Snyder F. Arachidonate-containing triacylglycerols: biosynthesis and a lipolytic mechanism for the release and transfer of arachidonate to phospholipids in HL-60 cells. Biochim Biophys Acta 1993; 1170:275-282.
12. Waite M. The Phospholipases. In: Hanahan DJ, ed. Handbook of Lipid Research. New York and London; Plenum Press, 1987.
13. Billah MM, Anthes JC. The regulation and cellular function of phosphatidylcholine hydrolysis. Biochem J 1990; 269:281-291.
14. Dennis EA, Rhee SG, Billah MM, Hannun Y. Role of phospholipases in generating lipid second messengers in signal transduction. FASEB J 1991; 5:2068-2077.
15. Mayer RJ, Marshall LA. New insights on mammalian phospholipase A_2(s); comparison of arachidonoyl-selective and non-selective enzymes. FASEB J 1993; 7:339-348.
16. Glaser KB, Mobilio D, Chang JY, Senko N. Phospholipase A_2 enzymes: regulation and inhibition. Trends Pharmacol Sci 1993; 14:92-98.

17. Dennis EA. Diversity of group types, regulation, and function of phospholipase A_2. J Biol Chem 1994; 269:13057-13060.
18. Clark JD, Milona N, JL Knopf. Purification of a 110-kilodalton cytosolic phospholipase A_2 from the human monocytic cell line U937. Proc Natl Acad Sci USA 1990; 87:7708-7712.
19. Gronich JH, Bonventre JV, Nemenoff RA. Purification of a high-molecular-mass form of phospholipase A_2 from rat kidney activated at physiological calcium concentrations. Biochem J 1990; 271:37-43.
20. Kramer RM, Roberts EF, Manetta J, Putnam JE. The Ca^{2+}-sensitive cytosolic phospholipase A_2 is a 100-kDa protein in human monoblast U937 cells. J Biol Chem 1991; 266:5268-5272.
21. Wijkander J, Sundler R. An 100-kDa arachidonate-mobilizing phospholipase A_2 in mouse spleen and the macrophage cell line J774. Purification, substrate interaction and phosphorylation by protein kinase C. Eur J Biochem 1991; 202:873-880.
22. Fujimori Y, Murakami M, Kim DK, Hara S, Takayama K, Kudo I, Inoue K. Immunochemical detection of arachidonoyl-preferential phospholipase A_2. J Biochem 1992; 111:54-60.
23. Omitted in proofs.
24. Sharp JD, White DL, Chiou XG et al. Molecular cloning and expression of human Ca^{2+}-sensitive cytosolic phospholipase A_2. J Biol Chem 1991; 266:14850-14853.
25. Lin LL, Wartmann M, Lin AY, Knopf JL, Seth A, Davis RJ. $cPLA_2$ is phosphorylated and activated by MAP kinase. Cell 1993; 72:269-278.
26. Nemenoff RA, Winitz S, Qian NX, Van Putten V, Johnson GL, Heasley LE. Phosphorylation and activation of a high molecular weight form of phospholipase A_2 by p42 microtubule-associated protein 2 kinase and protein kinase C. J Biol Chem 1993; 268:1960-1964.
27. Peppelenbosch MP, Qiu R-G, de Vries-Smits AMM, Tertoolen LGJ, de Laat SW, McCormick F, Hall A, Symons MH, Bos JL. Rac mediates growth-factor induced arachidonic acid release. Cell 1995 81:849-8546.
28. Hunter T. Protein kinases and phosphatases: the Yin and Yang of protein phosphorylation and signaling. Cell 1995; 80:225-236.
29. Gronchi V. Ricerche sulla lecitinasi 'A' del pancreas; estrazione dell'enzima e sua azione in vitro. Sperimentale 1932; 90:223-230.
30. Hanasaki K, Arita H. Characterization of a high-affinity binding site for pancreatic-type phospholipase A_2 in the rat. Its cellular and tissue distribution. J Biol Chem 1992; 267:6414-6420.
31. Arita H, Hanasaki K, Nakano T, Oka S, Teraoka H, Matsumoto K. Novel proliferative effects of phospholipase A_2 in Swiss 3T3 cells via specific binding site. J Biol Chem 1991; 266:19139-19141.

32. Kanemasa T, Hanasaki K, Arita H. Migration of vascular smooth muscle cells by phospholipase A_2 via specific binding sites. Biochim Biophys Acta 1992; 1125:210-214.
33. Nakajima M, Hanasaki K, Ueda M, Arita H. Effect of pancreatic-type phospholipase A_2 on isolated porcine cerebral arteries via its specific binding sites. FEBS Letters 1992; 309:261-264.
34. Forst S, Weiss J, Elsbach P, Maragamore JM, Reardon I, Heinrikson RL. Structural and functional properties of a phospholipase A_2 purified from an inflammatory exudate. Biochemistry 1986; 25:8381-8385.
35. Hayakawa M, Horigome K, Kudo I, Tomita M, Nojima S, Inoue K. Amino acid composition and NH_2-terminal amino acid sequence of rat platelet secretory phospholipase A_2. J Biochem 1987; 101:1311-1314.
36. Kanda A, Ono T, Yoshida N, Tojo H, Okamoto M. The primary structure of a membrane-associated phospholipase A_2 from human spleen. Biochem Biophys Res Comm 1989; 163:42-48.
37. Seilhamer JJ, Pruzanski W, Vadas P, Plant S, Miller JA, Kloss J, Johnson LK. Cloning and recombinant expression of phospholipase A_2 present in rheumatoid arthritic synovial fluid. J Biol Chem 1989; 264:5335-5338.
38. Aarsman AJ, de Jong JGN, Arnoldussen E, Neijs FW, van Wassenaar PD, Van den Bosch H. Immunoaffinity purification, partial sequence and subcellular fractionation of rat liver phospholipase A_2. J Biol Chem 1989; 264:10008-10014.
39. Van Schaik RHN, Verhoeven NM, Neijs FW, Aarsman AJ, Van den Bosch H. Cloning of the cDNA coding for 14 kDa group II phospholipase A_2 from rat liver. Biochim Biophys Acta 1993; 1169:1-11.
40. Murakami M, Kudo I, Natori Y, Inoue K. Immunochemical detection of 'platelet type' phospholipase A_2 in the rat. Biochim Biophys Acta 1990; 1043:34-42.
41. Kramer RM, Hession C, Johansen B, Hayes G, McGray P, Chow EP, Tizard R, Pepinski RB. Structure and properties of a human non-pancreatic phospholipase A_2. J Biol Chem 1989; 264: 5768-5775.
42. Murakami M, Kudo I, Inoue K. In vivo release and clearance of rat platelet phospholipase A_2. Biochim Biophys Acta 1989; 1005:270-276.
43. Gramzow M, Schröder HC, Fritsche U, Kurelec B, Robitzki A, Zimmermann H, Friese K, Kreuter MH, Müller WEG. Role of phospholipase A_2 in the stimulation of sponge cells proliferation by homologous lectin. Cell 1989; 59:939-948.
44. Lambeau G, Ancian P, Barhanin J, Lazdunski M. Cloning and expression of a membrane receptor for secretory phospholipase A_2. J Biol Chem 1994; 269:1575-1578.

45. Ishizaki J, Hanasaki K, Higashino K, Kishino J, Kikuchi N, Ohara O, Arita H. Molecular cloning of pancreatic group I phospholipase A_2 receptor. J Biol Chem 1994; 269:5897-5904.
46. Pernas P, Masliah J, Olivier JL, Salvat C, Rybkine T, Bereziat G. Type II phospholipase A_2 recombinant overexpression enhances stimulated arachidonic acid release. Biochem Biophys Res Comm 19991; 178:1298-1305.
47. Mounier C, Faili A, Vargaftig BB, Bon C, Hatmi M. Secretory phospholipase A_2 is not required for arachidonic acid liberation during platelet activation. Eur J Biochem 1993; 216:169-175.
48. Hazen SL, Stuppy RJ, Gross RW. Purification and characterization of canine myocardial cytosolic phospholipase A_2. J Biol Chem 1990; 265:10622-10630.
49. Ackermann EJ, Kempner ES, Dennis EA. Ca^{2+}-independent cytosolic phospholipase A_2 from macrophage-like P388D_1 cells. Isolation and characterization. J Biol Chem 1994; 269:9227-9233.
50. Balsinde J, Bianco I, Ackermann EJ, Conde-Frieboes K, Dennis EA. Inhibition of calcium-independent phospholipase A_2 prevents arachidonic acid incorporation and phospholipid remodeling in P3881D_1 macrophages. Proc Natl Acad Sci USA 1995; 92: 8527-8531.
51. Gross RW, Ramanadham S, Kruska KK, Han X, Turk J. Rat and human pancreatic islet cells contain a calcium ion independent phospholipase A_2 activity selective for hydrolysis of arachidonate which is stimulated by adenosine trisphosphate and is specifically localized to islet β-cells. Biochemistry 1993; 32:327-336.
52. Clark MA, Conway TM, Shorr RGL, Crooke ST. Identification and isolation of a mammalian protein which is antigenically and functionally related to the phospholipase A_2 stimulatory peptide mellitin. J Biol Chem 1987; 262:4402-4406.
53. Clark MA, Özgur LE, Conway TM, Dispoto J, Crooke ST, Bomalaski JS. Cloning of a phospholipase A_2-activating protein. Proc Natl Acad Sci USA 1991; 88:5418-5422.
54. Steiner MR, Bomalaski JS, Clark MA. Responses of purified phospholipases A_2 to phospholipase A_2-activating protein (PLAP) and mellitin. Biochim Biophys Acta 1993; 1166:124-130.
55. Clark MA, Chen MJ, Crooke ST, Bomalaski JS. Tumor necrosis factor (cachectin) induces phospholipase A_2 activity and synthesis of phospholipase A_2-activating protein in endothelial cells. Biochem J 1988; 259: 125-132.
56. Majerus PW, Connolly TM, Deckmyn H, Ross TS, Bross TE, Ishii H, Bansal VS, Wilson DB. The metabolism of phosphoinositide-derived messenger molecules. Science 1986; 234:1519-1536.

57. Bell RL, Kennerly DA, Stanford N, Majerus PW. Diglyceride lipase: a pathway for arachidonate release from human platelets. Proc Natl Acad Sci USA 1979; 76:3238-3241.
58. Allen AC, Gammon CM, Ousley AH, McCarthy KD, Morell P. Bradykinin stimulates arachidonic acid release through the sequential actions of an *sn*-1 diacylglycerol lipase and a monoacylglycerol lipase. J Neurochem 1992; 58:1130-1139.
59. Exton JH. Phosphatidylcholine breakdown and signal transduction. Biochim Biophys Acta 1994; 1212:26-42.
60. Klein J, Chalifa V, Liscovitch M, Löffelholz K. Role of phospholipase D activation in nervous system physiology and pathophysiology. J Neurochem 1995; 65:1445-1455.
61. Thomson FJ, Clark MA. Purification of a phosphatidic acid hydrolyzing phospholipase A_2 from rat brain. Biochem J 1995; 306:305-309.
62. Moolenar WH. Lysophosphatidic acid signalling. Current Opinion Cell Biol 1995; 7:203-210.
63. Miyake A, Yamamoto H, Kubota E, Hamaguchi K, Kouda A, Honda K, Kawashima H. Suppression of inflammatory responses to 12-O-tetradecanoyl-phorbol-13-acetate and carrageenin by YM-26734, a selective inhibitor of extracellular group II phospholipase A_2. Br J Pharmacol 1993; 110:447-453.
64. Beaton HG, Bennion C, Connolly S et al. Discovery of new nonphospholipid inhibitors of the secretory phospholipases A_2. J Med Chem 1994; 37:557-559.
65. Street IP, Lin HK, Laliberté F et al. Slow- and tight-binding inhibitors of the 85-kDa human phospholipase A_2. Biochemistry 1993; 32:5935-5940.
66. Barbour SE, Dennis EA. Antisense inhibition of group II phospholipase A_2 expression blocks the production of prostaglandin E_2 by $P388D_1$ cells. J Biol Chem 1993; 268:21875-21882.
67. Bennet CF, Chiang MY, Wilson-Lingardo L, Wyatt JR. Sequence specific inhibition of human type II phospholipase A_2 enzyme activity by phosphorothioate oligonucleotides. Nucleic Acid Research 1994; 22:3202-3209.
68. Elmore MA, Carrier MJ, Daniels RH, Hill ME, Finnen MJ. Antisense to secretory type II phospholipase A_2 inhibits fMLP-induced arachidonic acid release in differentiated HL60's. Biochemical Society Transactions 1994; 22:S136.
69. Vial D, Piomelli D. Dopamine D_2 receptors potentiate arachidonate release via activation of cytosolic, arachidonate specific phospholipase A_2. J Neurochem 1995; 64:2765-2772.
70. Gupta SK, Diez E, Heasley L, Osawa S, Johnson GL. A G protein mutant that inhibits thrombin and purinergic receptor activation of phospholipase A_2. Science 1990; 249:662-666.

71. Lin LL, Lin AY, Knopf JL. Cytosolic phospholipase A_2 is coupled to hormonally regulated release of arachidonic acid. Proc Natl Acad Sci USA 1992; 89:6147-6151.
72. Balsinde J, Barbour SE, Bianco I, Dennis EA. Arachidonic acid mobilization in P3881D_1 macrophages is controlled by two distinct phospholipase A_2 enzymes. Proc Natl Acad Sci USA 1994; 91:11060-11064.
73. Pfeilschifter J, Schalwijk C, Brine VA, Van den Bosch H. Cytokine-stimulated secretion of group II phospholipase A_2 by rat mesangial cells. J Clin Invest 1993; 92:2516-2523.
74. Hoeck WG, Ramesha CS, Chang DJ, Fan N, Heller RA. Cytoplasmic phospholipase A_2 activity and gene expression are stimulated by tumor necrosis factor: dexamethasone blocks the induced synthesis. Proc Natl Acad Sci USA 1993; 90:4475-4479.
75. Domin J, Rozengurt E. Platelet-derived growth factor stimulates a biphasic mobilization of arachidonic acid in Swiss 3T3 cells. The role of phospholipase A_2. J Biol Chem 1993; 268:8927-8934.
76. Jelsema CL. Light activation of phospholipase A_2 in rod outer segments of bovine retina and its modulation by GTP-binding proteins. J Biol Chem 1987; 262:163-168.
77. Jelsema CL, Axelrod J. Stimulation of phospholipase A_2 activity in bovine rod outer segments by the $\beta\gamma$ subunits of transducin and its modulation by GTP-binding proteins. Proc Natl Acad Sci USA 1987; 84:3625-3627.
78. Traiffort E, Ruat M, Arrang JM, Leurs R, Piomelli D, Schwartz JC. Expression of a cloned rat histamine H_2 receptor mediating inhibition of arachidonate release and activation of cAMP accumulation. Proc Natl Acad Sci USA 1992; 89:2649-2653.
79. Piomelli D. Arachidonic acid in cell signaling. Current Opinion Cell Biol 1993; 5:274-280.
80. Piomelli D, Pilon C, Giros B, Sokoloff P, Martres MP, Schwartz JC. Dopamine activation of the arachidonic acid cascade as a basis for D1/D2 receptor synergism. Nature 1991; 353:164-167.
81. Di Marzo V, Vial D, Sokoloff P, Schwartz JC, Piomelli D. Selection of alternative G_i-mediated signaling pathways at the dopamine D_2 receptor by protein kinase C. J Neurosci 1993; 13:4846-4853.
82. Clark D, White FJ. D1 receptors—the search for a function: a critical evaluation of the D_1/D_2 dopamine receptor classification and its functional implications. Synapse 1987; 1:347-388.
83. Jacob M, Weech PK, Salesse C. Presence of a light-independent phospoholipase A_2 in bovine retina but not in rod outer segments. J Biol Chem 1996; 271:19209-19218.

CHAPTER 3

Before Metabolism: Arachidonate as an Intracellular Second Messenger

Although our views on what an intracellular second messenger should be and how it should act have changed since cyclic AMP was discovered in the late 1950s, the set of experimental criteria originally proposed by Earl Sutherland to define a second messenger molecule are still applicable today. These criteria, listed in Table 3.1, are based on the definition of a second messenger as an endogenous molecule whose role is to transduce within a target cell the actions of a hormone, a neurotransmitter or any other extracellular signaling substance (i.e., a 'first messenger').

Table 3.1. Criteria for a second messenger substance

1. The enzymatic machinery necessary for the biosynthesis of the putative second messenger should be present in cells
2. Activating first messenger receptors should elicit the accumulation of the putative second messenger
3. A mechanism should exist for the rapid inactivation of the putative second messenger
4. Exogenous administration of the putative second messenger should mimic the biological response resulting from stimulating first messenger receptors
5. Drugs that prevent formation of the putative second messenger should also prevent the first messenger response

From what we have seen in the preceding chapter, arachidonic acid meets at least two of these criteria: like 'classical' second messengers, e.g., cyclic AMP or inositoltrisphosphate, the cellular amounts of free arachidonate are finely tuned by a balance between receptor-dependent formation and enzymatic disposition. The question we need to address now is whether free arachidonate can *transduce* the actions of extracellular messenger molecules by interacting with intracellular target proteins. In the present chapter, I shall describe two circumstances in which non-esterified arachidonate has been shown to modulate protein functions directly, i.e., without undergoing further metabolism: these are the regulation of membrane ion channels and the stimulation of the calcium- and phospholipid-dependent protein kinase, PKC.

Before I recapitulate these examples, it seems appropriate to discuss briefly two points which, though often neglected, are pertinent to the cellular roles of non-esterified arachidonate. We have seen that extracellular messengers mobilize free arachidonate, but how much of it do they mobilize? In other words, what concentration can free arachidonate reach in a stimulated cell? Answering this question case by case (i.e., cell by cell, and stimulus by stimulus) is essential to determine whether an observed biological action of exogenous arachidonate is likely to mimic that of an evoked physiological response. Moreover, arachidonate is poorly soluble in water: given the alternative, it will partition in lipid membranes 10^4 times more than in an aqueous phase. Thus, its actual concentration and its cellular sites of action will be dictated not only by the mass produced upon stimulation, but also by the presence of mechanisms of intracellular transport between membrane compartments. Is free arachidonate transported from one location to another within a cell, and if so, how? In the next two sections, I shall review first some of the evidence on the quantitative aspect of arachidonate mobilization, and then I shall outline the possible roles of lipid-binding proteins in intracellular arachidonate transfer.

HOW MUCH FREE ARACHIDONATE IS PRODUCED IN CELLS?

Measuring the concentration of free arachidonate in cells raises several technical problems. Methods based on radioisotope labeling are of little use because they do not provide the needed quantitative information on the *mass* of free arachidonate produced. Ultraviolet (UV) spectroscopy or immunological methods, which are invaluable to quantify many arachidonic acid metabolites (particularly when these methods are used in conjunction with high performance liquid chromatography, HPLC, see chapter 4), are precluded to non-metabolized arachidonate because of its poor UV absorbance and low immunogenicity.

How may one circumvent such difficulties? Unfortunately, there is no simple recipe, but two possible approaches are outlined in Figure 3.1. In either approach, free arachidonate must be first extracted from tissues (typically by organic solvent extraction) and partially purified from other lipid components by thin-layer chromatography (TLC). The TLC fraction containing free arachidonate (and other free fatty acids) is then subjected to an appropriate

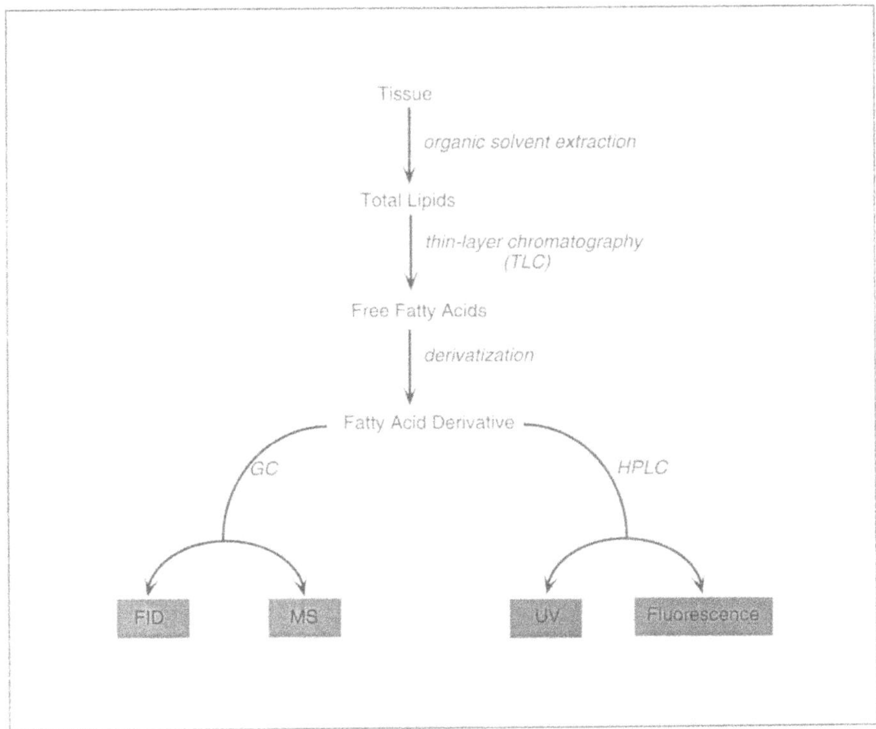

Fig. 3.1. Common approaches to the measurement of non-esterified arachidonate in animal tissues. Abbreviations used are: FID, flame ionization detection; GC, gas-liquid chromatography; HPLC, high performance liquid chromatography; MS, mass spectrometry; UV, ultraviolet spectrometry.

chemical derivatization reaction, which will depend on the choice of the subsequent analytical step.

To make free fatty acids amenable to analysis by gas chromatography (GC), it is necessary to convert them to completely nonpolar derivatives, such as the methylesters. These are readily prepared with commercially available reagents, and are resolved with great efficiency by GC, especially when using capillary columns. They can be identified unequivocally by combining information on their GC retention times with their mass spectral characteristics. An example of this approach is provided in Figure 3.2 which shows, in the upper panel, a GC chromatogram of the methylester derivative of arachidonic acid and, in the lower panel, its electron-impact mass spectrum: in the latter, the presence of an extensive and complex fragmentation, typical of the methylesters of polyunsaturated fatty acids, is noticeable.

If quantity is a limiting factor in the analysis, as it is often the case, one may increase the sensitivity of detection by selecting from the mass spectrum of the fatty acid derivative one or two diagnostic ions of reasonable abundance and focus the mass spectrometer on those, a technique called 'selected ion monitoring' (SIM). (For instance, one may choose the ion of m/z 175 from the spectrum of methylarachidonate shown in Fig. 3.2.) Alternatively, detection by flame ionization (FID) may be routinely used when quantities of the analyte are not limiting.

Although arachidonate and other fatty acids may be analyzed by HPLC without derivatization (for example, by linking the HPLC column to a light-scattering detector), derivatives prepared with either UV-absorbing or fluorescent tags are often preferred because they can be detected with much greater sensitivity. The reader interested in these technical issues is referred to several comprehensive books that have recently been published.[1-3]

GC/MS coupled to SIM detection has been used by John Turk and coworkers (at the Washington University School of Medicine in Saint Louis) to quantify the accumulation of free fatty acids which occurs in pancreatic islets, when these are stimulated with concentrations of D-glucose that promote insulin release from β cells.[4] Each unstimulated islet was found to contain approximately 0.5 pmole of free arachidonate, and this level was significantly increased when the islets were superfused with a solution containing D-glucose. Since no arachidonate could be detected in the medium superfusing the islets, Turk and colleagues concluded that the fatty acid remained associated with cells; assuming that the free fatty acid is distributed evenly through an intracellular islet volume of 2-4 nL, they calculated that islets would contain, after D-glucose stimulation, a concentration of free arachidonate comprised between 38 and 75 μM. Though very high, this value probably underestimates the actual concentration of free fatty acid *within membranes*. Also, as we shall see below, this calculated value is well within the concentrations at which exogenous arachidonate is

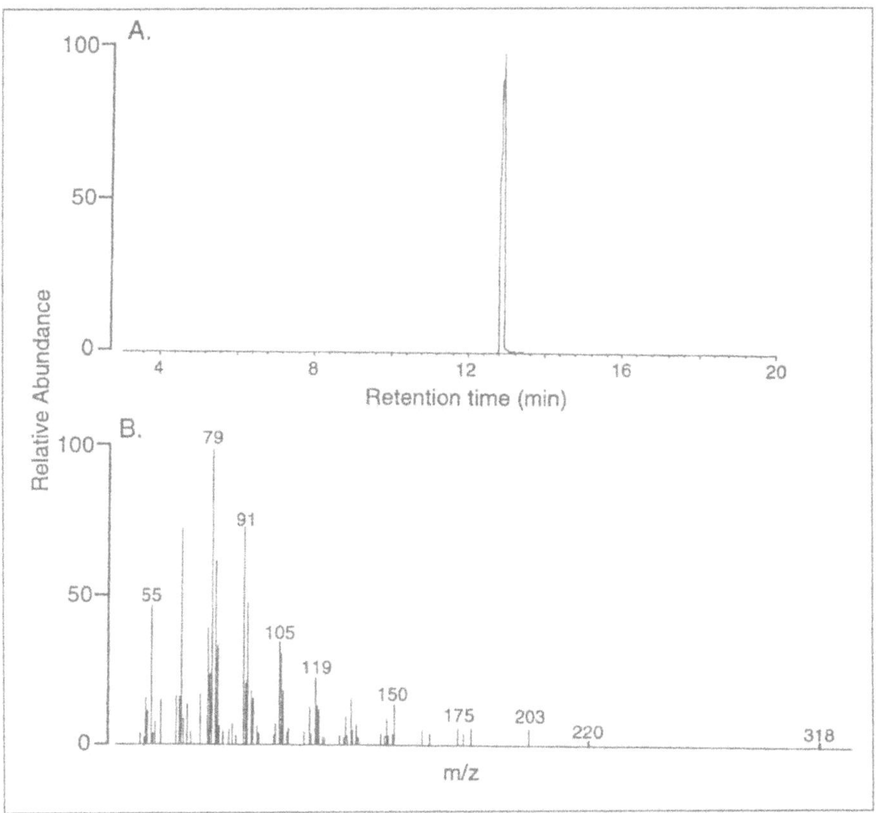

Fig. 3.2. Analysis of methylarachidonate. The upper panel shows the gas chromatographic behavior of standard methylarachidonate on a capillary column of 5% crosslinked phenylmethylsilicone (Hewlett-Packard), eluted with He_2 as carrier gas and using a temperature gradient from 150°C to 280°C at 8°C /min. The lower panel shows the electron impact mass spectrum of methylarachidonate obtained using a Hewlett-Packard 5972 series mass spectrometer.

maximally effective in modulating the activities of membrane ion channels and PKC. If concentrations similar to those measured in pancreatic islets were to be found in other cells, this would be indeed a strong argument in favor of a second messenger role of non-esterified arachidonate. It is unfortunate that only a few studies have addressed thus far this important issue using physiological stimuli.

HOW IS ARACHIDONATE TRANSPORTED WITHIN CELLS?

In the aqueous environment surrounding cells, the movement of water-insoluble compounds is facilitated by their interaction with lipid-binding proteins such as albumin, very low density lipoproteins and lipocalins.[5-7] These proteins are structurally heterogeneous, but share the ability to bind avidly to fatty acids (as well as to other endogenous or exogenous hydrophobic molecules) and to carry them through the extracellular fluid to their various cellular targets. As shown schematically in Figure 3.3, albumin and lipocalins bind non-covalently to free fatty acids, whereas very low density lipoprotein and chylomicrons bind to triacylglycerols. Fatty acids can be mobilized from triacylglycerols by the action of a lipoprotein lipase located on the cell surface (Fig. 3.3).[8] Do analogous mechanisms facilitate *intracellular* transfer of free arachidonate? Although we cannot give a definite answer to this question yet, evidence indicates that such mechanisms may exist.

A family of cytosolic proteins have been identified which bind long-chain fatty acids, including arachidonic acid, with high affinity (dissociation constants are in the range of 0.1-2 μM).[9,10] These fatty acid-binding proteins (FABP) are very abundant (for example, they are present in wet liver tissue at a concentration of 200 nmol/g), ubiquitous, tissue-specific polypeptides with 127-133 amino acid residues and molecular masses comprised between 14 and 16 kDa. A large number of FABP have been identified and several have been characterized. While they share a variable degree of homology at the level of their primary structure (sometimes as low as 20%), they are all very similar in their tertiary, tridimensional structure. X-ray crystallographic studies have demonstrated that all FABP are organized into two short α-helices and 10 antiparallel β-strands, which lie in two almost orthogonal β-sheets.

Interestingly, this tridimensional conformation is similar to that of extracellular lipocalins, whose amino acid sequences share no similarity with those of cytosolic FABP.[9,11]

Because of their binding to long-chain fatty acids and of their highly regulated expression in tissues, FABP are thought to play a central role in the uptake and transport of fatty acids, and possibly in other aspects of the biological actions and metabolism of these lipids. Indeed, FABP have been shown, in vitro, to transfer

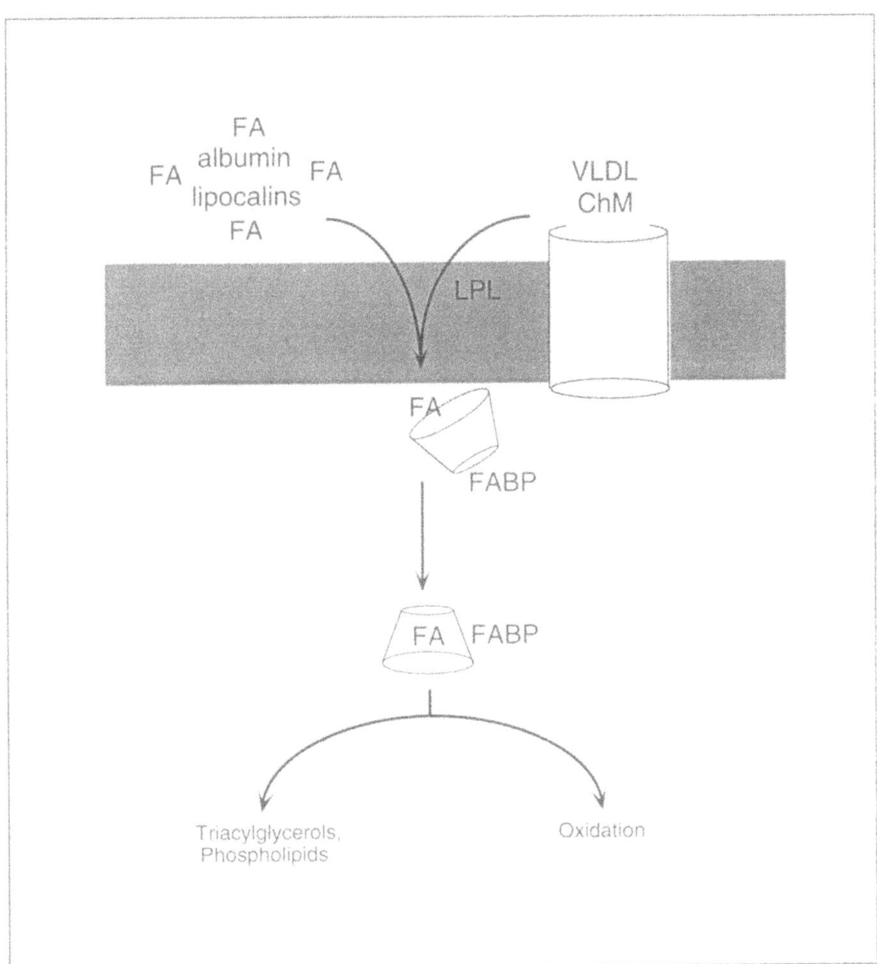

Fig. 3.3. Possible roles of intracellular fatty acid-binding proteins (FABP) in fatty acid metabolism. Fatty acids are delivered to cells by extracellular fatty acid-binding proteins (albumin, lipocalins), by very low-density lipoproteins (VLDL) or by chylomicrons (ChM). After internalization, the FABP may bind the non-esterified fatty acid, and may facilitate their transport to appropriate subcellular sites where metabolism takes place.

fatty acids from one membrane (either natural or artificial) to another.[12] It would be important to learn more about their functions in intact cells: to determine if FABP act exclusively as "solubilizing agents",[11] or rather, if they "help" the fatty acids in subtler ways (Fig. 3.4). For example, do they participate in the effects, described below, that non-esterified arachidonate exerts on ion channels and on protein kinase activities? A possibility is that FABP may act by presenting the fatty acids to their target proteins. Alternatively, the FABP-fatty acid complex might operate as an *activated*

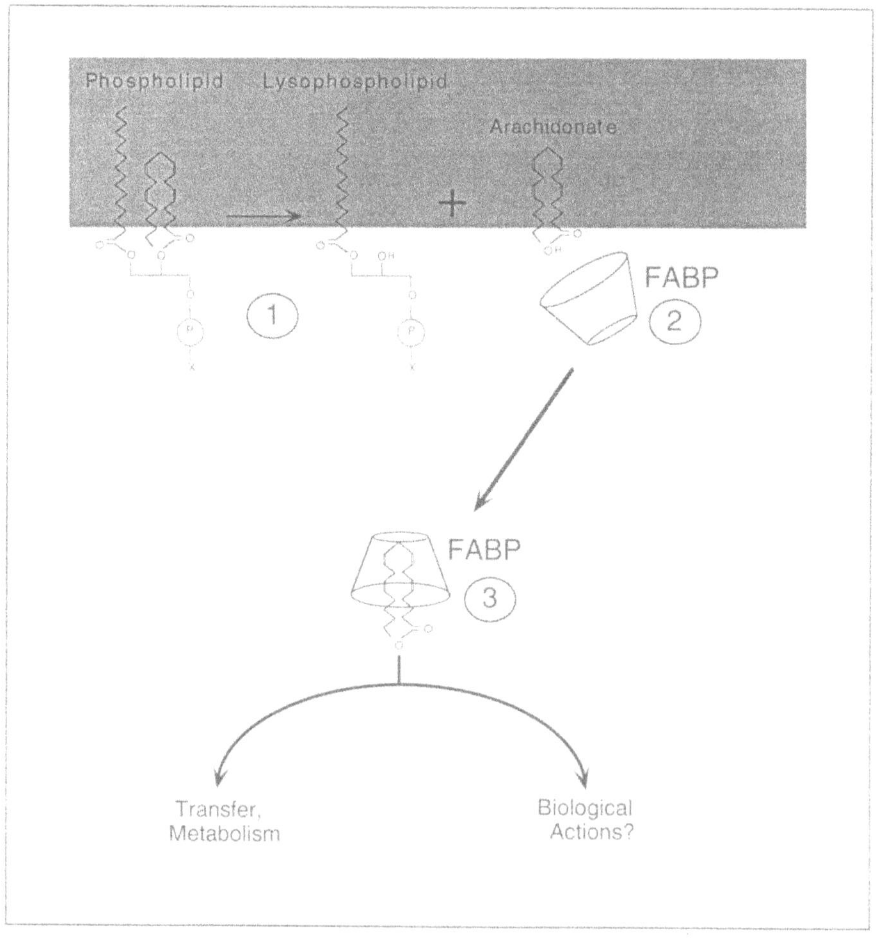

Fig. 3.4. Hypothetical roles of FABP in intracellular arachidonate trafficking. Arachidonate, released by phospholipase activities (1), binds to intracellular FABP (2). The arachidonate/FABP complex may mediate transfer, metabolism and possibly biological actions of non-esterified arachidonate within cells (3).

form of FABP, much in the same way as the calcium-calmodulin complex is an activated form of calmodulin, which transduces many of the intracellular actions of calcium. Whatever the answers to these questions, it appears that FABP deserve a greater attention for their possible roles in intracellular lipid signaling than they have received thus far.

DIRECT MODULATION OF ION CHANNELS BY ARACHIDONATE

Now that we have qualified, at least in first approximation, what we mean by endogenous free arachidonate levels, we may turn to our first example of the biological effects produced by the administration of this fatty acid as a pharmacological agent.

Exogenous arachidonate modifies the activity of many types of membrane ion channels in excitable cells. It does so through three basic mechanisms, which are listed in Figure 3.5. The simplest and most direct of these mechanisms involves the interaction of the non-metabolized fatty acid with an ion channel molecule or with the lipid environment that surrounds it: I will refer to such mechanism as 'direct modulation'. Indirect forms of modulation require, instead, either arachidonate activation of a protein kinase (usually PKC; see following section), or the enzymatic conversion of arachidonate into one of its metabolites: prostaglandins, hydroperoxides, leukotrienes, etc (see chapter 4). In the present section, I will concentrate on the direct modulation of ion channels: I will describe first a series of studies on the regulation of potassium channel activity by fatty acids in the gastric smooth muscle of the toad (*Bufo marinus*) and then summarize work done on the regulation of glutamate receptor-channels in rat brain neurons.

The direct regulation of potassium channels by fatty acids has been discovered by the laboratory of Joshua J. Singer (at the University of Massachusetts Medical School), using patch-clamping techniques.[13,14] Patch-clamp allows one to measure the activity of an individual ion channel while controlling the chemical composition of both the intracellular and extracellular milieu: a polished glass micropipette filled with an electrolyte solution is put in contact with the plasma membrane of a cell, forming a very high-resistance electrical seal (Fig. 3.6). A patch of the membrane is

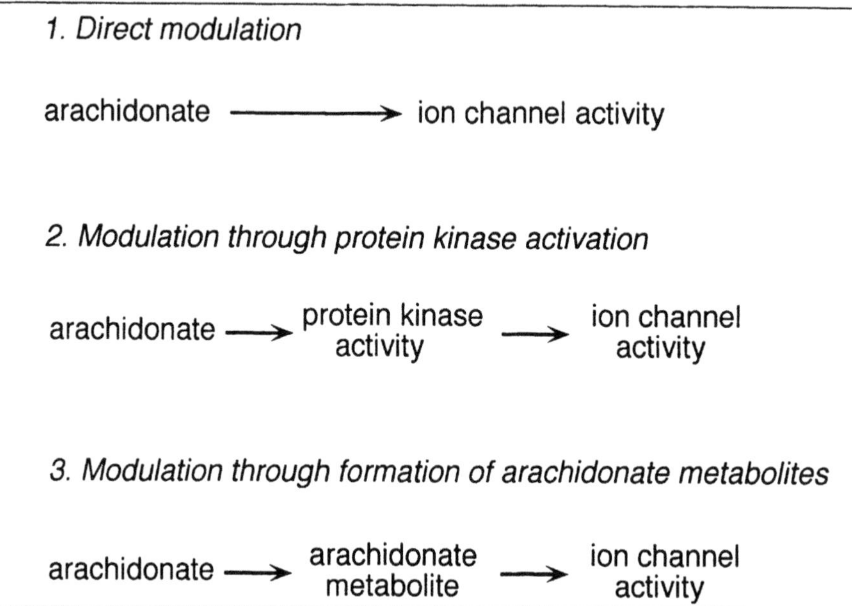

Fig. 3.5. Different modalities of ion channel regulation by non-esterified arachidonate in animal cells.

then aspirated into the micropipette and excised from the surface of the cell in such a way that its intracellular aspect may either face the bathing solution (*inside-out* configuration) or the solution within the electrode (*outside-out*). Because of the high electrical resistance established between cell membrane and glass micropipette, extremely small currents such as those generated by the flux of ions across a channel pore can be measured reliably. After appropriate electric amplification, the opening of a single channel appears as an all-or-none current step: its unitary *conductance* is represented by the height of the step and its *open probability* by the time spent in the open state (Fig. 3.6).

Singer and colleagues studied the effect of non-esterified arachidonate on potassium channels by dissolving the fatty acid in the bathing medium and perfusing it on the inside of the smooth muscle membrane (in the *inside-out* configuration) or on its outside (in the *outside-out* configuration). In either case, they found that application of micromolar concentrations of the fatty acid would cause the activation of a potassium-selective channel (Fig. 3.6). This effect was not mediated by the formation of eicosanoids because, first, it was not prevented by drugs that inhibit arachidonate metabolism and, second, it was mimicked by fatty acids that are not substrates for cycloxygenase, lipoxygenase or cytochrome P_{450} (e.g., myristic acid, 14:0). Moreover, the effect did not involve protein phosphorylation, as it could be produced in the absence of ATP, GTP or cyclic nucleotides.

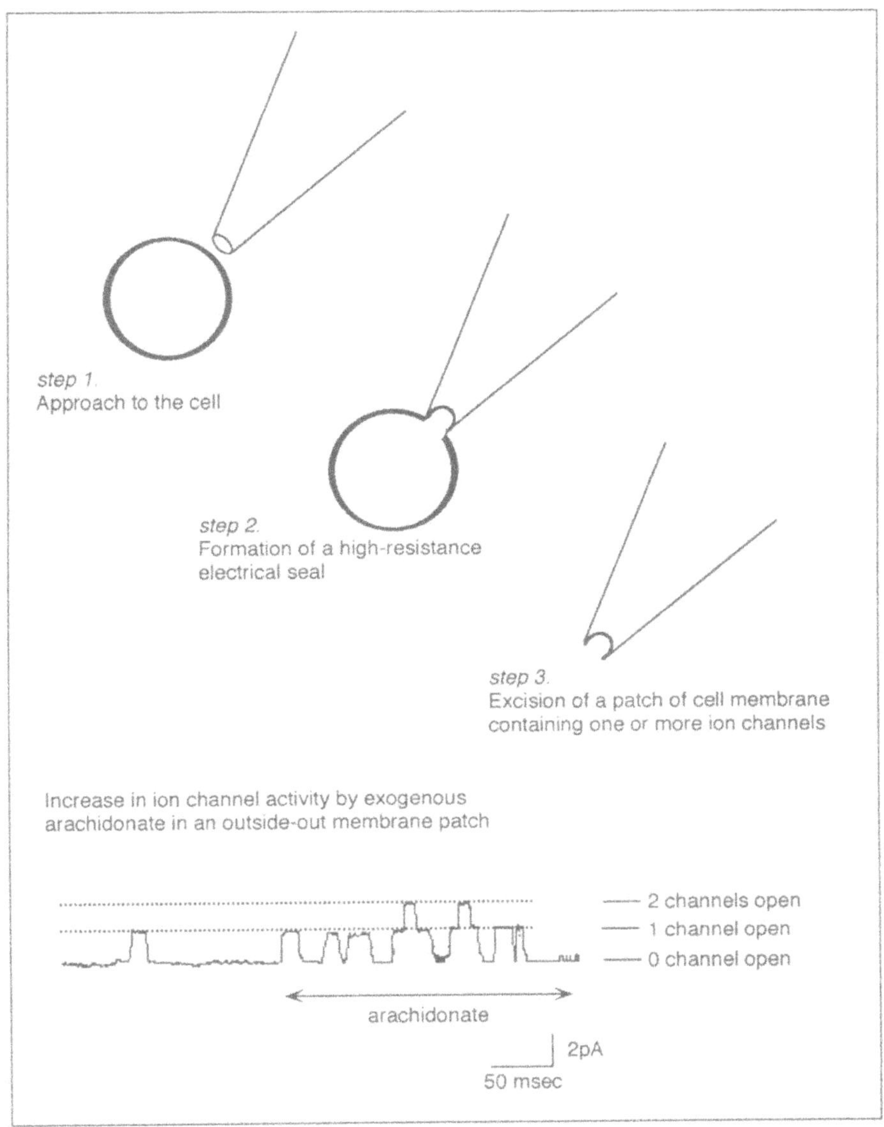

Fig. 3.6. Arachidonate activation of potassium channels in vertebrate smooth muscle cells: analysis by patch clamp. Scheme drawn from experiments reported in refs. 13 and 14.

Beside arachidonate, a large number of hydrophobic molecules bearing a negatively charged group were found to increase potassium channel activity; by contrast, positively charged lipids were found to close the potassium channel. The structures of some lipid molecules shown to activate potassium channels in toad smooth

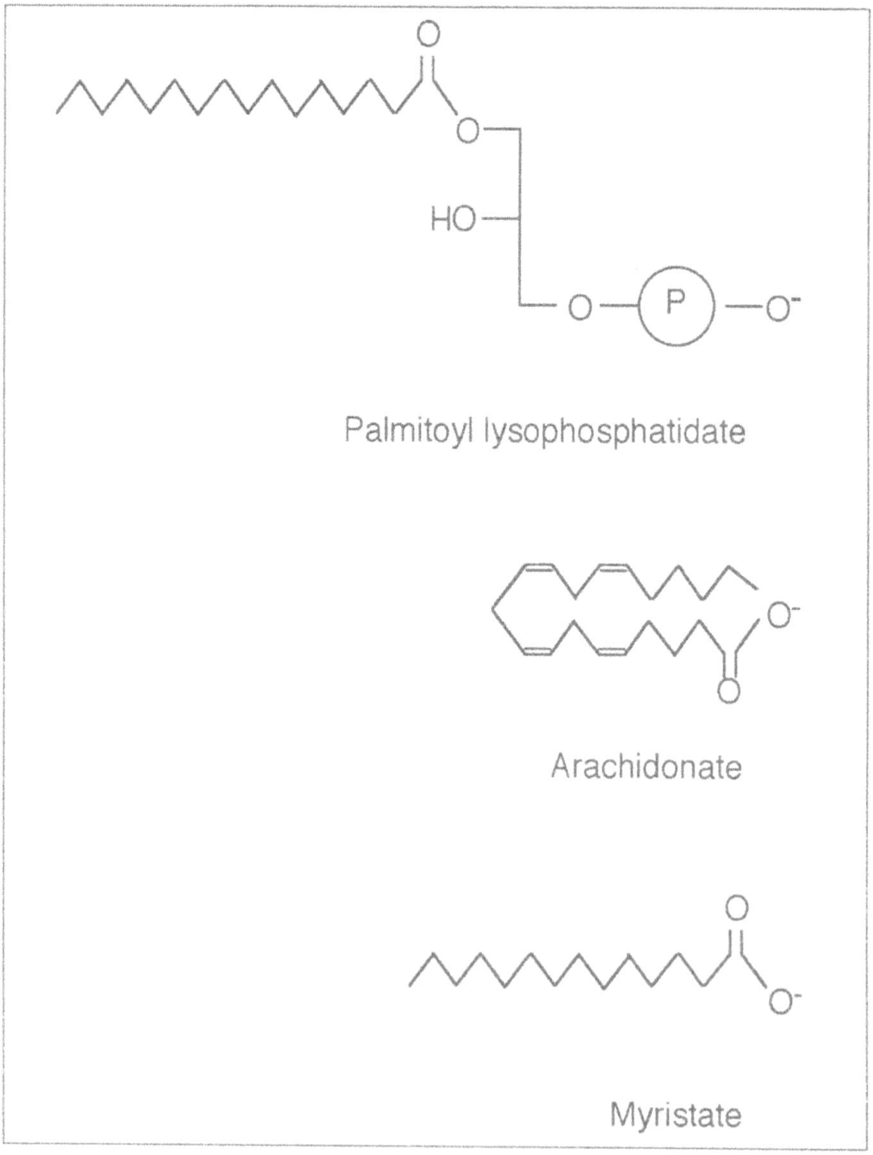

Fig. 3.7. Chemical structures of various lipid molecules that activate potassium channels in toad smooth muscle.[13,14]

muscle are depicted in Figure 3.7. Several of them, such as fatty acids and lysophosphatidic acid, are naturally-occurring substances which can be produced in a receptor-dependent manner, suggesting that they may act as endogenous potassium channel activators. Because activation of potassium channels in excitable cells tends to hyperpolarize the membrane potential and to decrease amplitude and duration of the action potential, the physiological role of these lipids would be to decrease the cell's excitability. This possibility is supported by the fact that non-esterified fatty acids are also potent in decreasing the activity of voltage-dependent calcium channels.[15] Functionally, this action would be synergistic with potassium channel activation and may result in a concerted inhibitory effect on the electrical activity of the cell.

Let us consider now our second example of ion channel regulation by non-esterified arachidonate—the potentiation of glutamate receptor currents in mammalian brain neurons. Glutamate, the major excitatory neurotransmitter in the brain, acts by binding to a heterogenous group of transmembrane receptors that are either coupled to second messengers (metabotropic receptors) or contain in their structure a pore that lets ions through when the receptor is stimulated (ionotropic receptors). Ionotropic receptors are further subdivided into two pharmacological classes: those that are sensitive to the glutamate receptor agonist, N-methyl-D-aspartate (NMDA-type), and those that are not (non-NMDA-type). The activity of NMDA glutamate-gated channels is governed not only by glutamate released from nerve terminals, but also by a small number of allosteric regulators present at the synapse. The amino acid, glycine, is one such molecule and evidence suggests that non-esterified arachidonate may be another one.

David Attwell and collaborators, at the University College in London, found that arachidonate increases, in cerebellar granule cells, the current flowing through activated NMDA channels, whereas it reduces the current through non-NMDA channels.[16] This potentiating effect appears to be produced by a direct interaction of the fatty acid with the channel molecule or with its lipid environment, as it is not prevented by inhibitors of arachidonic acid metabolism or protein kinase activity. Interestingly, the NMDA receptor contains a domain of 131 amino acid residues that is homologous to the presumed fatty acid-binding site of cytosolic FABP: it is reasonable to suppose, therefore, that binding of arachidonate to this domain of the receptor protein may underlie the fatty acid's ability to modulate ion channel activity.[17]

What may be the physiological significance of this modulation? A possible clue comes from the finding that, by activating NMDA receptors, glutamate causes arachidonate mobilization in brain neurons (see ref. 18 for review). This fact points to the hypothetical scenario shown in Figure 3.8, where NMDA receptors are proposed to serve as both the trigger and the target for non-esterified arachidonate. Activation of NMDA receptors raises cytosolic calcium levels, leading to increased arachidonate mobilization via activation of an as yet unknown phospholipase: if appropriate concentrations of the fatty acid are reached in relevant cell compartments, this increase would cause, in turn, a potentiation of the NMDA response. Such a feed-forward loop might participate

Figure 3.8

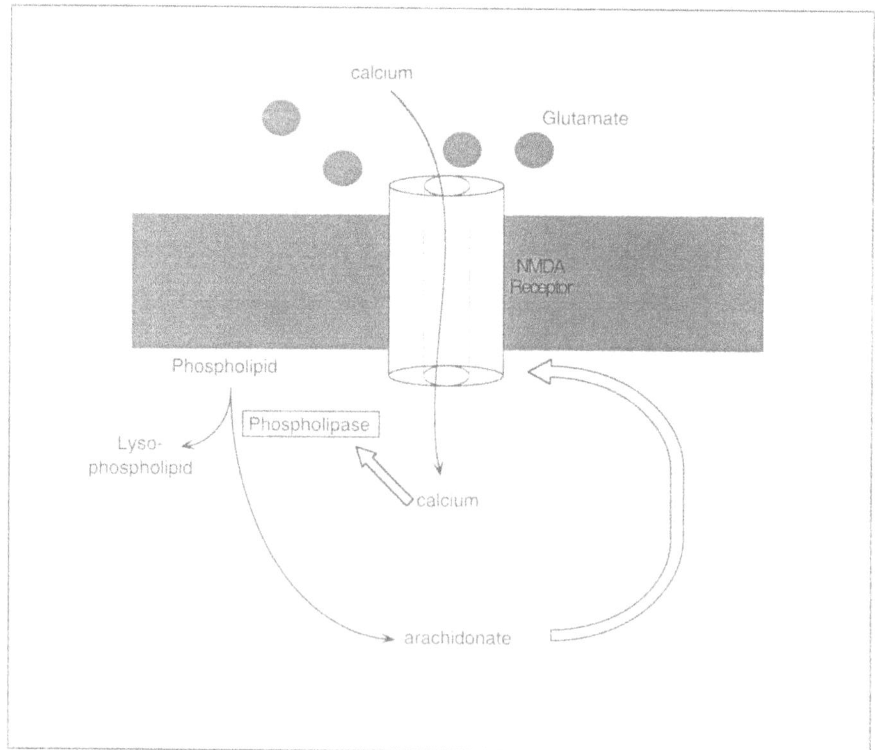

Fig. 3.8. Hypothetical model of the regulation of glutamate NMDA receptor channels by non-esterified arachidonate in neurons.

in long-lasting forms of synaptic potentiation that depend on the activation of NMDA receptor-channels.

In conclusion, non-esterified arachidonate shares with other natural lipids the ability to activate various membrane ion channels, some of which are listed in Table 3.2. The precise mechanisms through which these modulatory effects are brought about are not known yet, but they may involve a direct interaction of the lipids with a hydrophobic component located either within the channel molecule or in close association with it.

ARACHIDONATE MODULATION OF PROTEIN KINASE C

PKC is a family of protein kinase isozymes which are thought to be present in, and essential for, the functioning of virtually all

Table 3.2. Direct effects of non-esterified arachidonate on membrane ion channels

Direct effects of non-esterified arachidonate on membrane ion channels

ChannelType	*Effect*	*Concentration* (μM)
K^+ channels		
Outward rectifier		
Hippocampal pyramidal cells	Activation	10-100
Gastric smooth muscle	Activation	10-100
Transient outward		
Ventricular muscle	Activation	1-50
Vascular smooth muscle	Activation	20
Atrial and ventricular muscle	Activation	10
Cl^- channels		
Airway epithelium	Inhibition	1-25
NMDA receptor-channels		
Central neurons	Activation	0.5-50

cell types. This short section is of course not the appropriate place to review even cursorily the complex regulation and numerous biological roles of this protein kinase. It suffices to remind that most of the isoforms of PKC become activated in tissues when extracellular stimuli increment cytosolic calcium levels and stimulate the hydrolysis of membrane phospholipids, producing 1,2-DAG. A combination of calcium and 1,2-DAG or, depending on the isoform, 1,2-DAG alone, activates PKC by binding to specific control regions of the protein kinase molecule, resulting in the phosphorylation and functional modification of many protein substrates.[19] The fact that elevated cytosolic calcium and enhanced phospholipid hydrolysis are also associated with the mobilization of arachidonate and other unsaturated fatty acids has led Linda C. McPhail and her collaborators (at the Howard Hughes Medical Institute in Durham) to suggest that these lipids may participate in the control of PKC activity.[20] Working with detergent extracts of human neutrophils, they observed that unsaturated fatty acids stimulate PKC activity in a calcium-dependent manner and that this effect is potentiated by the concomitant application of 1,2-DAG. Subsequently, these findings have been confirmed and extended by several laboratories, and particularly by that of Yasutomi Nishizuka at Kobe University in Japan.[19,21] It is now widely accepted that unsaturated fatty acids (e.g., oleate, linoleate and arachidonate) enhance the 1,2-DAG-dependent activation of PKC and that this effect is particularly prominent at lower calcium concentrations (Table 3.3). As in the case of potassium channels, PKC activation is not mediated by metabolism of the fatty acid, and it is likely due to a direct interaction of the lipid with a putative regulatory element within the protein kinase structure. Unlike potassium channels, however, saturated fatty acids and other negatively charged lipids have no effect on PKC activity, indicating a greater selectivity of this element.

An interesting case of synaptic modulation that may depend upon arachidonate stimulation of PKC activity has been reported by Immaculada Herrero, José Sánchez-Prieto and their coworkers at the Universidad Complutense in Madrid.[22] Glutamatergic nerve terminals isolated from the rat brain express a metabotropic subtype of the glutamate receptor that is coupled to inositol phospholipid metabolism and PKC activation. Stimulation of nerve terminals with the metabotropic receptor agonist, *trans*-1-amino-cyclopentyl-1,3-dicarboxylate (*trans*-ACPD), results in enhanced glutamate release *only* if arachidonate (2 μM) is also supplied in the medium. That this facilitatory effect of arachidonate may be mediated by PKC is indicated by two findings: first, activating PKC with tumor-promoting phorbol esters produces a similar potentiation; second, inhibiting PKC with staurosporine prevents it. Thus, arachidonate mobilization and stimulation of presynaptic metabotropic receptors may act as a "coincidence detector": when both phenomena occur at the same synapse within a limited time-window, a greater release of glutamate from the nerve terminals may ensue. Under physiological conditions, free arachidonate may be produced by the stimulation of an additional receptor located on the nerve terminal, on the postsynaptic element or on adjacent neurons and glial cells (Fig. 3.9).

Table 3.3. Arachidonate-sensitive PKC isoforms in mammalian tissues

PKC Isoforms	*Arachidonate Sensitivity*	*Tissue Expression*
α	+	Ubiquitous
βI	+	Sparse
βII	+	Common
γ	+	Brain, Spinal Cord
δ	-	Ubiquitous
ε	+	Brain and Other Tissues
η	?	Skin, Lung, Heart
θ	?	Skeletal Muscle
ζ	+	Ubiquitous
λ	?	Ovary, Testes and other tissues

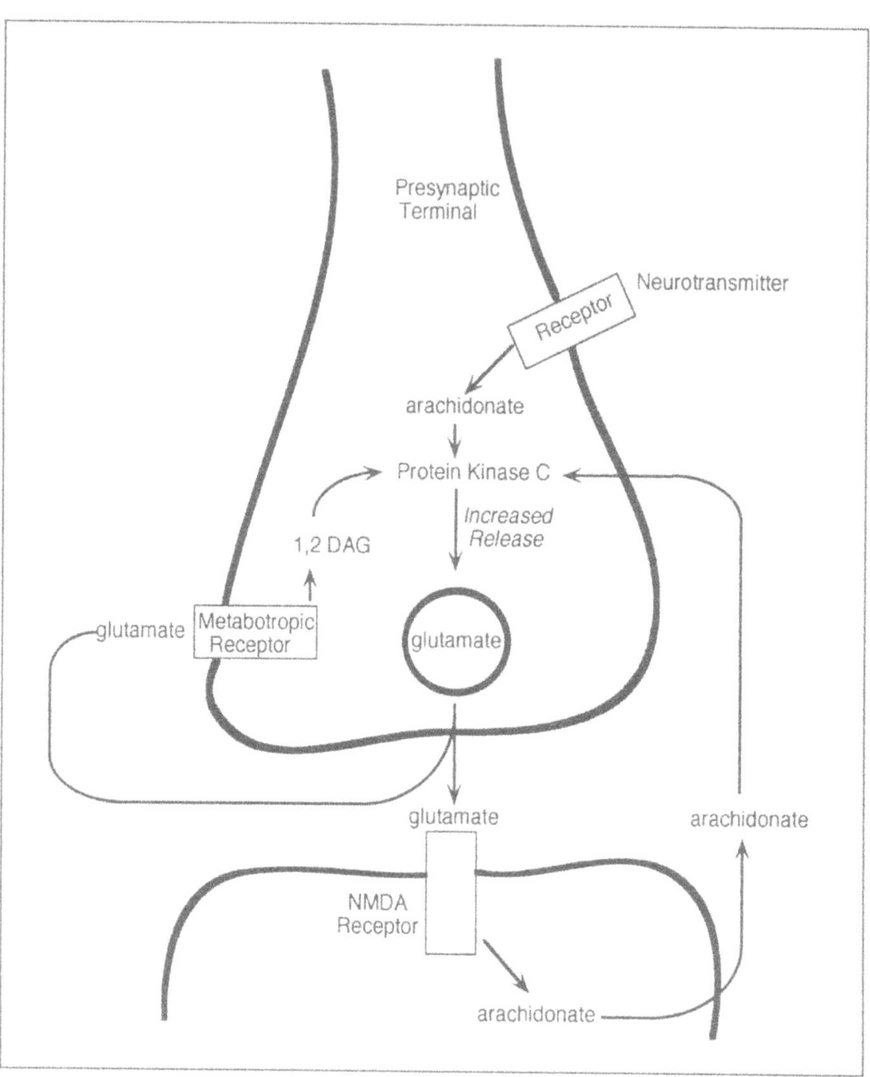

Fig. 3.9. Hypothetical model of protein kinase C activation by non-esterified arachidonate. Shown schematically are a presynaptic terminal juxtaposed to a postsynaptic spine. Glutamate, released by presynaptic electrical activity, binds to and stimulates postsynaptic NMDA receptors as well as presynaptic metabotropic receptors. NMDA receptors are linked to the release of free arachidonate[35], which in turn is thought to diffuse to the presynaptic terminal and to enhance 1,2-diacylglycerol (DAG)-stimulated protein kinase C activity. Alternatively, non esterified arachidonate may be produced presynaptically, by stimulation of phospholipase-linked neurotransmitter receptors. Based on results published in ref. 22.

The proposed synaptic roles of arachidonate extend to a fashionable model of synaptic plasticity and information storage in the mammalian brain: long-term potentiation (LTP) of synaptic transmission. LTP is believed to consist of two phases, induction and maintenance. Induction is initiated by the postsynaptic entry of calcium, which occurs through glutamate NMDA receptor channels. The mechanism of LTP maintenance is very controversial. Some investigators think that it may be produced at least partly by presynaptic mechanisms and, to bridge postsynaptic induction with presynaptic maintenance, they postulate the existence of a diffusible retrograde messenger.[23] That arachidonate may serve such function was suggested by several findings. First, as we have seen above, the stimulation of glutamate NMDA receptors elicits arachidonate mobilization in neurons.[18] Also, non-selective PLA_2 inhibitors prevent the induction of LTP, while application of arachidonate or other unsaturated fatty acids to hippocampal slices causes an activity-dependent enhancement of synaptic transmission.[24,25] The onset of this response is very slow, however, and does not match the time-course of 'authentic' LTP. While this discrepancy does not rule out a participation of arachidonate, it has played a role in diverting the attention of many researchers to other diffusible messengers, such as nitric oxide.[26]

PHYSIOLOGICAL ROLES?

Incomplete as it is, our sample testifies that much is known on what non-esterified arachidonate may do when it is administered to cells or to purified enzymes. The situation changes dramatically, though, when we come to consider the physiological meaning of these actions. On this issue the data are much scantier, and the results much more open to interpretation. Two experimental models in which the application of exogenous arachidonate has been combined fruitfully with the use of inhibitors of arachidonate mobilization may provide insight into the possible physiological functions of non-esterified arachidonate. These models are the mechanism of glucose-induced insulin release in pancreatic β cells, and the mechanism of potassium "leak channel" regulation by somatostatin in hippocampal pyramidal neurons.

We have seen above that D-glucose stimulates arachidonate mobilization in pancreatic islets, raising the levels of the non-esterified fatty acid to a calculated value of approximately 38-75 μM. The sugar is transported within the cell, phosphorylated to D-glucose-6-phosphate and funneled through the glycolytic pathway, producing adenosine trisphosphate (ATP). ATP initiates in turn a series of events leading, on the one hand, to insulin secretion, and on the other, to stimulation of a calcium-independent cytosolic PLA_2 activity, which mobilizes arachidonate from phospholipids (Fig. 3.10).[27-29] What is the function of arachidonate? To answer this question, John Turk and his collaborators have studied the effect of exogenous arachidonate on insulin secretion. They found that, when administered together with a depolarizing agent such as KCl, physiologically relevant concentrations of the free fatty acid (5-30 μM) facilitate insulin secretion from the islets.[4] This result suggests that non-esterified arachidonate may shift the threshold of insulin secretion toward higher (more negative) membrane potentials, possibly by affecting the voltage-dependence of membrane calcium channels. In support of this possibility, arachidonate was found to evoke an increment in β-cell cytosolic calcium, which occurred with the expected concentration-dependence (1-20 μM), and which was (a) reversible, (b) abolished by removal of extracellular calcium and (c) reduced by blockers of voltage-dependent calcium channels. Moreover, this effect was likely mediated by arachidonate itself rather than by a metabolite, as it was not affected by inhibitors of arachidonate metabolism.[30-32] Interestingly, cycloxygenase metabolites of arachidonic acid are also produced during D-glucose stimulation of pancreatic islets, but they are thought to participate in some form of *intercellular*, rather than intracellular, signaling function (Fig. 3.10). (12-Lipoxygenase products may also play a role,· see ref. 27 for discussion.)

Such branching into multiple messages that act on multiple targets is not uncommon with the arachidonic acid cascade, as it is also illustrated by our next example: the participation of arachidonate in the electrophysiological effects of the peptide, somatostatin. Somatostatin modulates the excitability of hippocampal pyramidal neurons by activating at least two types of potassium currents: a non-inactivating voltage-dependent current, called I_{KM} because it is decreased by cholinergic *m*uscarinic drugs, and a voltage-insensitive potassium current. Although the somatostatin receptor subtype responsible of mediating these responses is not known, pharmacological evidence indicates that both responses may be transduced by arachidonate mobilization. For instance, the administration of exogenous arachidonate mimics the somatostatin response, while PLA_2 inhibitors prevent it.[33] However, while the

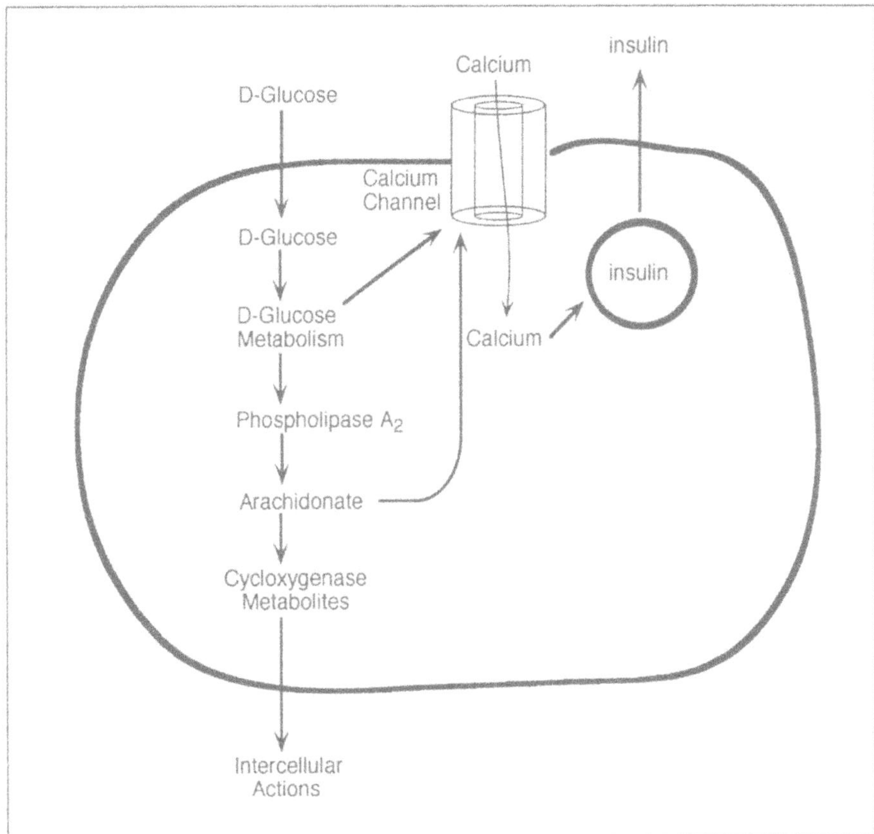

Fig. 3.10. Possible role of arachidonate in the amplification of insulin secretion from pancreatic islets. Intracellular glucose metabolism is thought to be linked to the stimulation of both calcium channels, which results in insulin secretion, and phospholipase A_2 activity, which increases the levels of non-esterified arachidonate. Arachidonate may in turn enhance insulin secretion by altering the voltage-dependence of calcium channels. In addition, arachidonate is converted into cyclooxygenase metabolites which may exit the cell, and modulate insulin secretion by binding to cell membrane receptors on adjacent β cells. Based on results published on refs. 27-32.

activating effect of somatostatin on I_{KM} is probably mediated by arachidonic acid metabolites produced via the 5-lipoxygenase pathway, the neuropeptide's effect on voltage-insensitive potassium conductance may be mediated by non-metabolized arachidonate.[34] This two-pronged transduction system, involving both non-esterified arachidonate and arachidonate metabolites, constitutes an appropriate preface to our next chapter, which will focus on the manifold metabolic transformations undergone by this fatty acid in cells.

REFERENCES

1. Hamilton RJ, Hamilton S. Lipid Analysis. A Practical Approach. Oxford, University Press, 1992.
2. Christie WW. Gas Chromatography and Lipids. A Practical Guide. Dundee, The Oily Press, 1989.
3. Christie WW. HPLC and Lipids. A Practical Guide. Oxford, Pergamon Press, 1987.
4. Wolf BA, Pasquale SM, Turk J. Free fatty acid accumulation in secretagogue-stimulated pancreatic islets and effects of arachidonate on depolarization-induced insulin secretion. Biochemistry 1991, 30:6372-6379.
5. Flower DR, North ACT, Attwood TK. Structure and sequence relationships in the lipocalins and related proteins. Protein Science 1993; 2:753-761.
6. Gachon, AMF. Lipocalins: do we taste with our tears? Trends Biochem Sci 1993; 18:206-207.
7. Nagata A, Suzuki Y, Igarashi M et al. Human brain prostaglandin D synthase has been evolutionarily differentiated from lipophylic-ligand carrier proteins. Proc Natl Acad Sci USA 1991; 88: 4020-4024.
8. Garfinkel AS, Schotz MC. Lipoprotein Lipase. In: Gotto AM, ed. Plasma Lipoproteins. Amsterdam, New York, Oxford: Elsevier, 1987:335-357.
9. Veerkamp JH, Peeters RA, Maatman RGHJ. Structural and functional features of different types of cytoplasmic fatty acid-binding proteins. Biochim Biophys Acta 1991; 1081:1-24.
10. Lalonde JM, Levenson MA, Roe JJ, Bernlorh DA, Banaszak LJ. Adipocyte lipid-binding protein complexed with arachidonic acid. Titration calorimetry and X-ray crystallographic studies. J Biol Chem 1994; 269:25339-25347.
11. Banaszak LJ, Winter N, Xu Z, Bernlorh DA, Cowan S, Jones TA. Lipid-binding proteins: a family of fatty acid and retinoid transport proteins. Advances in Protein Chemistry 1994; 45:89-151.

12. Peeters RA, Veerkamp JH, Demel RA. Are fatty acid-binding proteins involved in fatty acid transfer? Biochim Biophys Acta 1989; 1002:8-13.
13. Ordway RW, Singer JJ, Walsh Jr JV. Direct regulation of ion channels by fatty acids. Trends Neurosci 1991; 14:96-100.
14. Petrou S, Ordway RW, Hamilton JA, Walsh JV Jr, Singer JJ. Structural requirements for charged lipid molecules to directly increase or suppress K^+ channel activity in smooth muscle cells. Effects of fatty acids, lysophosphatidate, acyl coenzyme A and sphingosine. J Gen Physiol 1994; 103:471-486.
15. Meves H. Modulation of ion channels by arachidonic acid. Progress Neurobiol 1994; 43:175-186.
16. Miller B, Sarantis M, Traynelis SF, Attwell D. Potentiation of NMDA receptor currents by arachidonic acid. Nature 1992; 355:722-725.
17. Petrou S, Ordway RW, Singer JJ, Walsh JV Jr. A putative fatty acid-binding domain of the NMDA receptor. Trends Biochem Sci 1993; 18:41-42.
18. Piomelli D. Eicosanoids in synaptic transmission. Critical Rev Neurobiol 1994; 8:65-83.
19. Tanaka C, Nishizuka Y. The protein kinase C family for neuronal signaling. Annu Rev Neurosci 1994; 17:551-567.
20. McPhail LC, Clayton CC, Snyderman R. A potential second messenger role for unsaturated fatty acids: activation of Ca^{2+}-dependent protein kinase. Science 1984; 224:622-625.
21. Naor Z, Shearman MS, Kishimoto A, Nishizuka Y. Calcium-independent activation of hypothalamic type I protein kinase C by unsaturated fatty acids. Molecular Endocrinology 1988; 2: 1043-1048.
22. Herrero I, Miras-Portugal MT, Sánchez-Prieto J. Positive feedback of glutamate exocytosis by metabotropic presynaptic receptor stimulation. Nature 1992; 360:163-165.
23. Bliss TVP, Collingridge GL. A synaptic model of memory: long-term potentiation in the hippocampus. Nature 1993; 361:31-39.
24. Linden DJ, Sheu F-S, Murakami K, Routtemberg A. Enhancement of long-term potentiation by cis-unsaturated fatty acid: relation to protein kinase C and phospholipase A_2. J Neurosci 1992; 12:3601-3608.
25. Williams JH, Errington ML, Lynch MA, Bliss TVP. Arachidonic acid induces a long-term activity-dependent enhancement of synaptic transmission in the hippocampus. Nature 1989; 341:739-742.
26. Fazeli MS. Synaptic plasticity: on the trail of the retrograde messenger. Trends Neurosci 1992; 15:115-117.

27. Turk J, Gross RW, Ramanadham S. Amplification of insulin secretion by lipid messengers. Diabetes 1993; 42:367-374.
28. Gross RW, Ramanadham S, Kruszka KK, Han X, Turk J. Rat and human pancreatic islet cells contain a calcium ion independent phospholipase A_2 activity selective for hydrolysis of arachidonate which is stimulated by adenosine trisphosphate and is specifically localized to islet β cells. Biochemistry 1993; 32:327-336.
29. Ramanadham S, Wolf MJ, Jett PA, Gross RW, Turk J. Characterization of an ATP-stimulatable Ca^{2+}-independent phospholipase A_2 from clonal insulin-secreting HIT cells and rat pancreatic islets: a possible molecular component of the β-cell fuel sensor. Biochemistry 1994; 33:7442-7452.
30. Wolf BA, Turk J, Sherman WR, McDaniel ML. Intracellular Ca^{2+} mobilization by arachidonic acid. Comparison with *myo*-inositol 1,4,5 trisphosphate in isolated pancreatic islets. J Biol Chem 1986; 261:3501-3511.
31. Ramanadham S, Gross RW, Han X, Turk J. Inhibition of arachidonate release by segretagogue-stimulated pancreatic islets suppresses both insulin secretion and the rise in β-cell cytosolic calcium ion concentration. Biochemistry 1993; 32:337-346.
32. Ramanadham S, Gross RW, Turk J. Arachidonic acid induces an increase in the cytosolic calcium concentration in single pancreatic islet beta cells. Biochem Biophys Res Comm 1992; 184:647-653.
33. Schweitzer P, Madamba S, Siggins GR. Arachidonic acid metabolites as mediators of somatostatin-induced increase in neuronal M-current. Nature 1990; 346:464-467.
34. Schweitzer P, Madamba S, Champagnat J, Siggins GR. Somatostatin inhibition of hippocampal CA1 pyramidal neurons: mediation by arachidonic acid and its metabolites. J Neurosci 1993; 13:2033-2049.
35. Piomelli D. Arachidonic acid. In: Bloom FE, Kupfer DJ eds. Psychopharmacology: The Fourth Generation of Progress, New York: Raven Press 1995:595-607.

CHAPTER 4

Arachidonate Metabolism

We are about to discuss the complex series of biochemical transformations undergone by non-esterified arachidonate in cells and to outline the properties of the enzymatic systems involved in these reactions. As this task will oblige us to go into a considerable amount of factual detail, it may be helpful to begin with a general overview of arachidonate metabolism, whose landmarks are illustrated in Figure 4.1.

As we have seen in chapters 1 and 2, the metabolism of arachidonate begins with its mobilization from membrane phospholipids and consists mainly of an array of sequential oxidative reactions. Three main classes of oxygenase enzymes are implicated: prostaglandin H-synthase, also called cyclooxygenase; cytochrome P_{450} and a variety of different lipoxygenases.

Prostaglandin H-synthase catalyzes the first committed step in the conversion of arachidonate to the prostanoids, which consists in the oxygenation and ring closure of arachidonate to form prostaglandin H_2 (PGH_2). The prostanoids include the prostaglandins (PGE_2, $PGF_{2\alpha}$ and PGD_2), prostacyclin and thromboxane A_2. All prostanoids undergo further metabolism, which in some cases may be part of a degradative process (for example, the oxidation of PGE_2 to 15-keto PGE_2), while in others it may yield additional biologically active molecules (for example, the conversion of PGD_2 to 15-deoxy-$\Delta^{12,14}$ PGJ_2, a differentiating factor).

The first reaction catalyzed by all lipoxygenase enzymes is the addition of a hydroperoxide group to a select carbon atom in the arachidonate molecule, to form a chiral hydroperoxyeicosatetraenoic acid (HPETE). Thus, the primary product of mammalian 5-lipoxygenase is 5(S)-HPETE, that of sea-urchin 12-lipoxygenase,

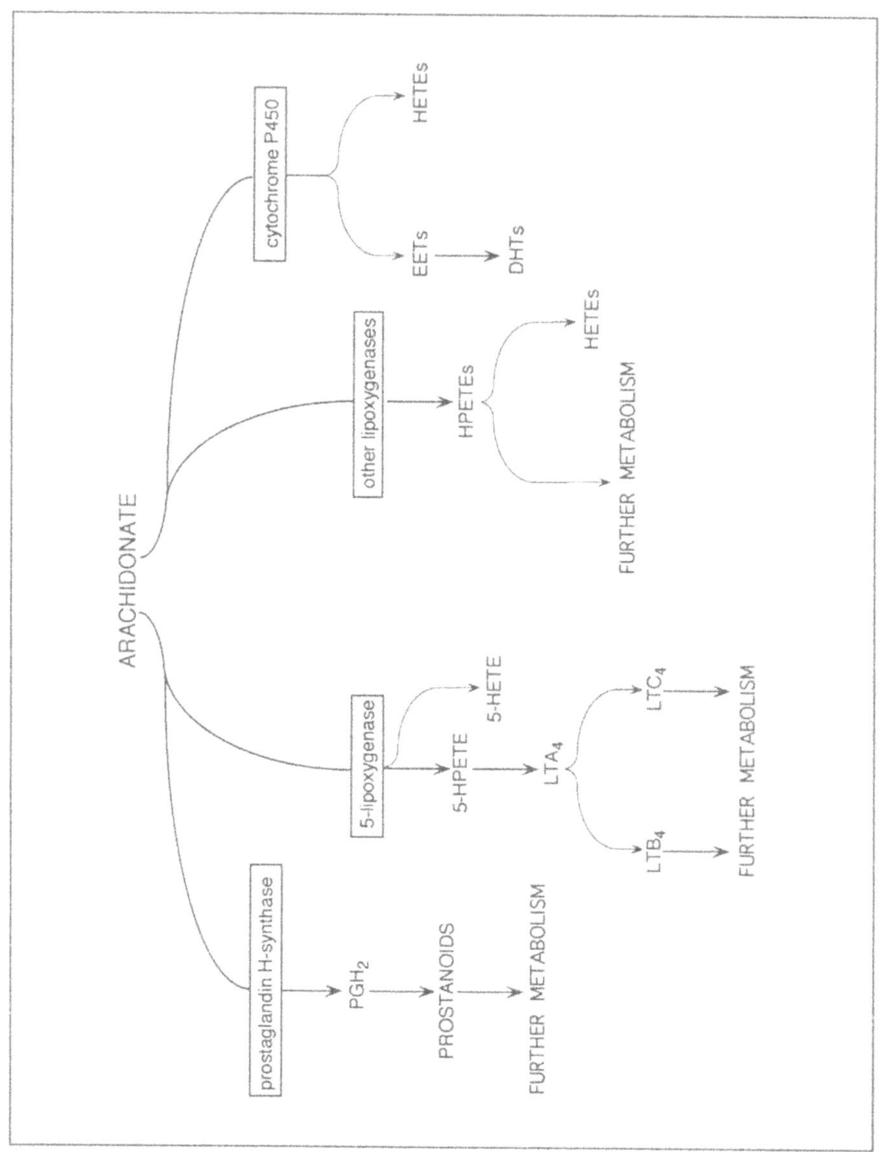

Fig. 4.1. Overview of the major pathways of arachidonate metabolism in vertebrate tissues. Abbreviations: dihydroxyeicosatrienoic acid, DHT; epoxyeicosatrienoic acid, EET; hydro(per)oxyeicosatetraenoic acid, H(P)ETE; leukotriene, LT; prostaglandin, PG.

12(R)-HPETE, etc. After this initial step, the metabolic routes followed by each HPETE are widely divergent, as shown, for example, by the different fates of 5(S)-HPETE and 12(S)-HPETE. 5(S)-HPETE is converted into the leukotrienes. These include leukotriene A_4 (LTA_4) a short-lived intermediate, as well as the more stable products, leukotriene B_4 (LTB_4), leukotriene C_4 (LTC_4) and leukotriene E_4 (LTE_4). 12(S)-HPETE gives rise, instead, to a group of metabolites that include hepoxilins and 12-hydroxyeicosatetraenoic acid, 12(S)-HETE (Fig. 4.1).

Cytochrome P_{450} catalyzes the monooxygenation of arachidonate to form a family of isomeric epoxyeicosatrienoic acids (EETs). These may be, in turn, transformed by the enzyme epoxide hydrolase to the corresponding diols, the dihydroxyeicosatrienoic acids (DHT). In addition, cytochrome P_{450} can also convert arachidonate into a family of nonchiral HETEs.

In the following sections I will briefly outline each of these pathways, describing the enzymes involved, their molecular structures, cellular localizations and pharmacological inhibition. Before turning to our main subject, however, it may be useful to give a general introduction to the analytical strategies that are most frequently employed to study the biochemistry of arachidonate metabolism.

ARACHIDONATE METABOLISM: AN OVERVIEW OF ANALYTICAL METHODS

After more than 30 years of intensive research on the eicosanoids, it is not surprising that a vast collection of analytical techniques are available to the investigator. Exhaustive reviews dealing with various methodological aspects of eicosanoid analysis may be found in the literature.[1,2] In the next pages, I will give a few examples of these methods and provide a series of select references to serve as an introductory guide.

Most biological samples contain complex mixtures of arachidonate metabolites, usually present at exceedingly low concentrations. This circumstance calls for some caution on the researcher's part. For example it is not advisable to measure eicosanoid levels in unfractionated samples, even when using specific immunochemical assays, because the chemical structures of the eicosanoids present in them may be similar enough to create problems of cross-reactivity. Therefore, the primary task of the analyst is to purify, as much as possible, the eicosanoid(s) of interest from one another as well as from contaminating material. This is usually achieved by a combination of extraction and chromatographic separation techniques, which are illustrated schematically in Figure 4.2.

A common way of extracting arachidonate metabolites from biological material is by using organic solvents. Typically, the aqueous samples are acidified, to protonate the eicosanoids' carboxylic

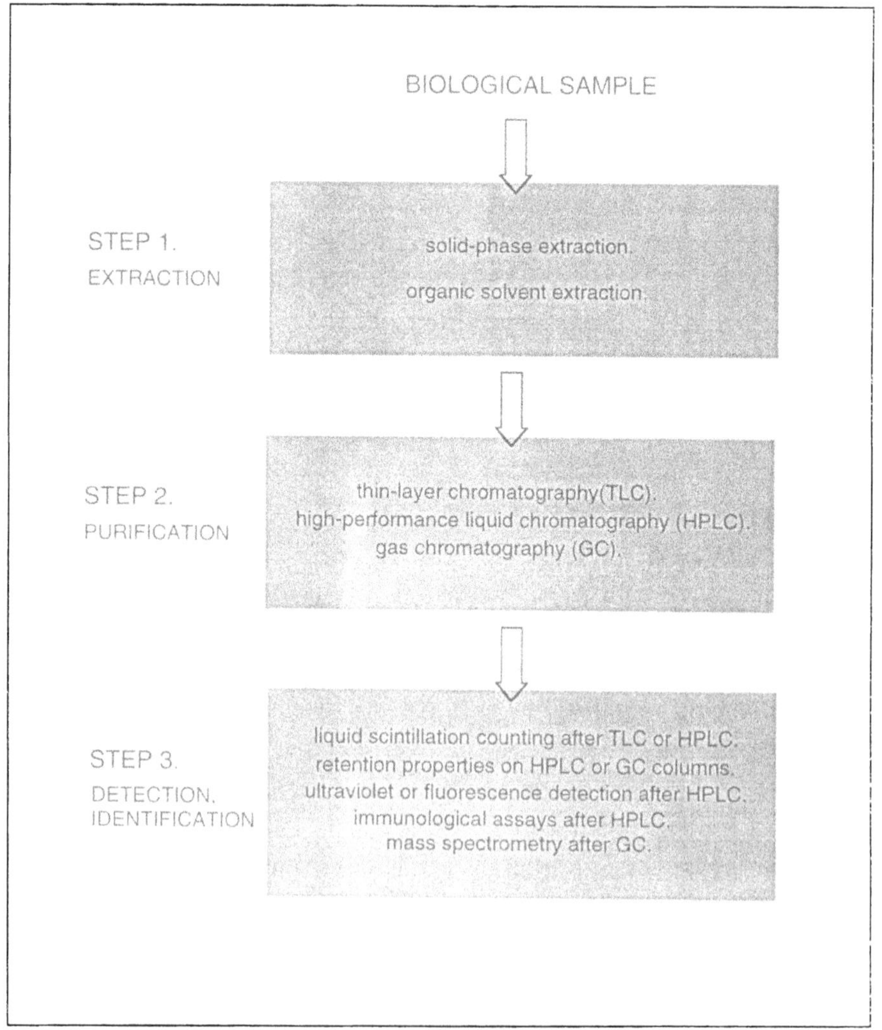

Fig. 4.2. Strategies in the biochemical analysis of the eicosanoids.

groups, and are extracted several times with diethyl ether, ethyl acetate or chloroform. An alternative procedure makes use of solid-phase extraction, often carried out on reversed-phase silica supports (e.g., octadecylsilyl silica, ODS).[1] In this case the aqueous samples are loaded directly onto the columns, and the arachidonate metabolites are eluted more or less selectively with appropriate combinations of various solvent systems.[1] In either case, the result is a mixture of substances (eicosanoids plus contaminants) which require further separations and analyses.

Thin-layer chromatography (TLC)[3,4] has played a pivotal role in the purification and structural characterization of the eicosanoids. For example, it was by using TLC coupled to analysis by gas chromatography/mass spectrometry that Mats Hamberg, Bengt Samuelsson and their collaborators at the Karolinska Institute first identified and isolated the thromboxanes.[5,6] TLC is still very useful, particularly when a positive identification of the analytes is not immediately required, as in sample work-up or in quantification of radioactively-labeled eicosanoids present in well-characterized biological samples.

Another major asset in the routine purification and analysis of the eicosanoids is high-performance liquid chromatography (HPLC).[7,8] In particular, reversed-phase HPLC has proven to be extremely effective, especially when combined with radioactive, spectroscopic or immunochemical detection methods.[1,2] For example, the combination of reversed-phase HPLC with ultraviolet absorbance spectroscopy was instrumental to the discovery of leukotrienes[9,10] and lipoxins,[11] and it is still widely used to detect these compounds as well as other ultraviolet-absorbing eicosanoids.[12]

Not all arachidonate metabolites have favorable absorbance properties, however. When this is the case, two additional techniques of high sensitivity may be used. The eicosanoids eluting from the HPLC column may be measured with the aid of an immunochemical assay (e.g., radioimmunoassay or enzyme-linked immunoassay).[13,14] Alternatively, these compounds may be chemically coupled to a chromophore before HPLC purification, and measured by ultraviolet or by fluorescence spectroscopy.[15] Direct analysis of the HPLC eluant by mass spectrometry (HPLC/MS) is also feasible, but its use has been hampered so far by high costs and by the limited availability of suitable HPLC/MS interfaces. This situation is likely to change in the near future.[16]

In the meantime, the method of choice to identify and measure low levels of arachidonate metabolites in complex biological samples remains gas chromatography/mass spectrometry (GC/MS).[2,16-18] After a series of clean-up steps—which may involve extraction, TLC and/or HPLC fractionation—the material is subjected to an appropriate reaction of derivatization. The derivatized compounds are then isolated by capillary GC and unambiguously identified by their GC retention times and mass spectral properties. When using electron-impact ionization, which ordinarily produces multiple fragments of the analyte, the sensitivity may be increased by choosing a few characteristic fragments and focusing the mass spectrometer on them (a technique known as selected ion monitoring, SIM). Precise quantification can be carried out by adding to the biological samples known quantities of standards containing stable isotopes (isotope dilution assays).[2] High-sensitivity detection may be also obtained by using negative-ion chemical ionization, which produces relatively simple spectra dominated by the analyte's molecular ion. An example of the application of these GC/MS approaches to the identification of a 12-lipoxygenase metabolite is illustrated in Figure 4.3.

At the beginning of this chapter, we have noted that the stereochemical composition of an eicosanoid may be indicative of the enzymatic system through which the compound is produced. Therefore, positive conclusions as to the enzymatic origin of a given eicosanoid can be best drawn when the product's stereochemistry is precisely defined. This task can be accomplished now with relative ease by HPLC on chiral-phase columns. This method and other methods used in the chiral analysis of arachidonate metabolites have been reviewed.[2]

ENZYMES OF PROSTANOID BIOSYNTHESIS

Prostaglandin Endoperoxide (PGH)-Synthase (Cyclooxygenase)

PGH-synthase catalyzes the conversion of arachidonate to PGH_2. This unstable intermediate is immediately transformed into prostaglandins, prostacyclin (PGI_2) and thromboxane A_2 (TXA_2) through the action of specific synthases.

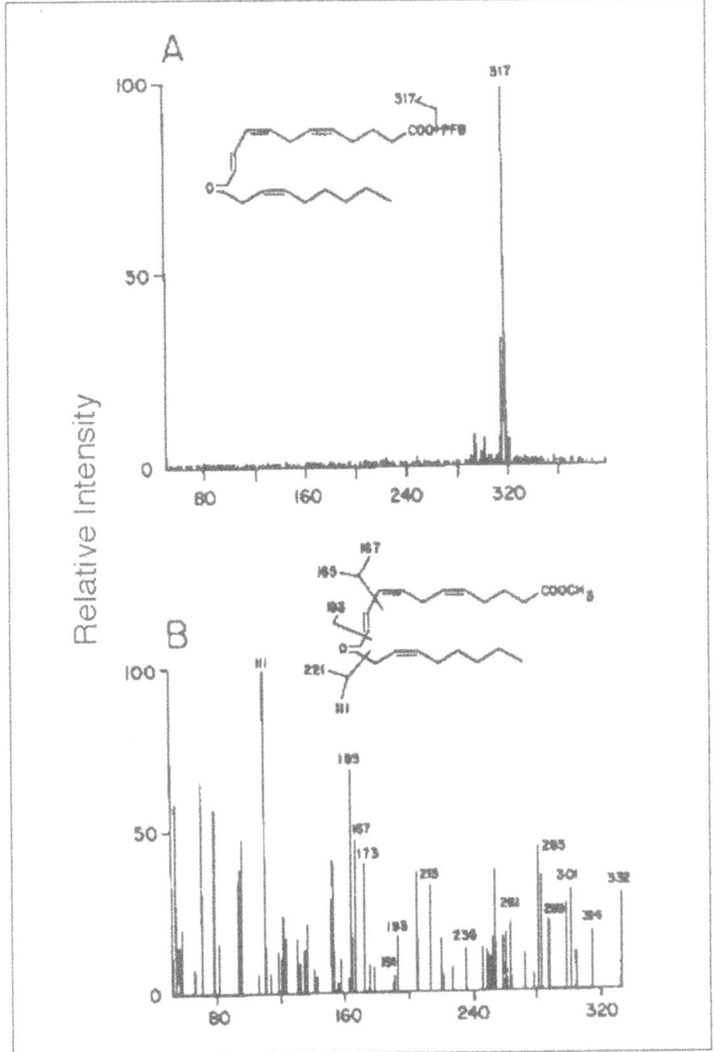

Fig. 4.3. Application of chemical-ionization and electron-impact ionization mass spectrometry to the identification of the 12-lipoxygenase product, 12-ketoeicosatetraenoic acid. Neural tissue from the marine mollusk, Aplysia californica, *was incubated for 30 minutes in the presence of exogenous arachidonate (50 μM). The incubates were extracted with organic solvent, and fractionated by HPLC. HPLC fractions containing 12-ketoeicosatetraenoic acid were brought to dryness, subjected to appropriate derivatization reactions, and analyzed by GC/MS using either negative-ion chemical ionization (top) or electron-impact ionization (bottom). The relatively simple spectrum obtained by chemical ionization provides information on the molecular mass of the compound. The electron-impact spectrum supports this information (a molecular ion of m/z 332 is visible) and provides additional elements for the unambiguous identification of the analyte. (Reproduced with permission from Piomelli D et al. J Biol Chem 1988; 263:16591-16596.)*

In mammals, there are two PGH-synthase isozymes with similar kinetic properties. They are encoded by distinct genes and differ in their regulation of expression and tissue distribution. PGH-synthase 1 (cyclooxygenase 1) has been purified from bovine and ovine seminal vesicles, and complementary DNAs encoding it have been cloned from sheep, mouse and human tissues.[19-26] PGH-synthase 1 is a heme-containing, integral membrane protein located primarily in the endoplasmic reticulum but also present in the plasmalemma and in the nuclear membrane. Its quaternary structure was elucidated by X-ray crystallography and consists of two identical and interconnected monomers.[20,27] PGH-synthase 1 is constitutively expressed in many tissues, where it is thought to be involved in cell-cell signaling and tissue homeostasis. By contrast, PGH-synthase 2 (cyclooxygenase 2) is an inducible enzyme that is normally absent from most cells (with the exception of brain, testes and macula densa cells of the kidney). Expression of PGH-synthase 2 is induced by mitogens and by inflammatory stimuli, suggesting that its activity may be responsible for the formation of prostanoids implicated in mitogenesis and inflammation.[28-33]

Both PGH-synthase 1 and PGH-synthase 2 carry out two sequential and distinct reactions (reviewed in refs. 34 and 35). First, a cyclooxygenase activity catalyzes the incorporation of two oxygen molecules into arachidonate, resulting in the formation of PGG_2 (15-hydroperoxy-9α,11α-peroxido-prostaenoic acid). Next, a peroxidase activity catalyzes the reduction of PGG_2 to PGH_2 (Fig. 4.4).

Cyclooxygenase Activity

The best substrates for the cyclooxygenase activity of PGH-synthase are arachidonate (K_M = 2-10 μM) and 8,11,14-eicosatrienoate. These fatty acids are both present in tissues, but arachidonate is more abundant and it is likely to be the most common substrate in vivo. By contrast, ω-3 polyunsaturated fatty acids, such as eicosapentaenoate (20:5 $\Delta^{5,8,11,14,17}$) or docosahexaenoate (22:6 $\Delta^{4,7,10,13,16,19}$), are, in general, poor substrates or even inhibitors of cyclooxygenase activity. (In ω-3 fatty acids the last double bond in the aliphatic chain is three carbon atoms away from the ω carbon; in ω-6 fatty acids, six carbon atoms away.)

X-ray diffraction studies of PGH-synthase 1 from sheep seminal vesicles have shown that the cyclooxygenase active site consists of an elongated hydrophobic channel extending from the external aspect of the membrane-binding domain toward the center of each PGH-synthase 1 monomer. This arrangement is illustrated schematically in Figure 4.5. (For a more precise stereo view of PGH-synthase 1, see ref. 27.) Two amino acid residues appear to be important in the catalytic process. These are Tyr 385, located at the top of the hydrophobic channel and close to a haem group, and Arg 120 located midway up the channel. It is hypothesized that arachidonate adopts a bent conformation within the active site, with the carboxylate group interacting with the positive charge of Arg 120, and the carbon 13 of its aliphatic chain located in the vicinity of Tyr 385. From a functional standpoint, Arg 120 may participate in docking the fatty acid to the enzyme's active site, while Tyr 385 may cooperate with the haem group in removing

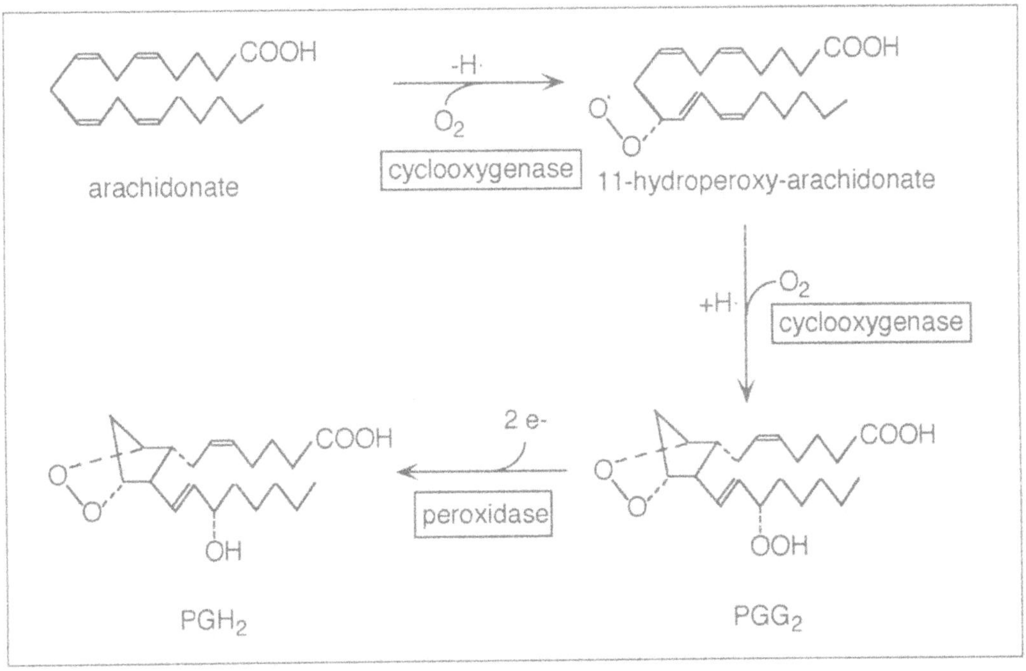

Fig. 4.4. Reactions catalyzed by PGH-synthase. The cyclooxygenase activity of PGH-synthase catalyzes the incorporation of two molecules of oxygen into arachidonate, producing PGG_2. This hydroperoxide is rapidly reduced by the peroxidase activity of PGH-synthase yielding PGH_2, the immediate precursor of prostaglandins, prostacyclin and thromboxane A_2.

one of the hydrogen atoms bound to carbon 13 (the pro-S hydrogen, Fig. 4.6). The resulting arachidonate radical can be then attacked at carbon 11 by an oxygen molecule, yielding the 11-peroxy radical, which cycles into a 9,11 endoperoxide. This intermediate is attacked by a second molecule of oxygen, producing PGG_2 (Fig. 4.6). It has been pointed out that, despite this apparently complex series of molecular events, the role of the cyclooxygenase active site may be in fact quite simple: to fix arachidonate in a curved conformation and allow for the stereospecific removal of the pro-S hydrogen atom at carbon 13.[28]

A third amino acid residue present within the cycloxygenase active site is also noteworthy. The Ser 530 residue, found just beneath Tyr 385 (Fig. 4.5), is irreversibly acetylated by aspirin, the non-steroidal anti-inflammatory drug that acts by inhibiting cyclooxygenase activity.[36,37] The location of Ser 530 suggests that acetylation of this residue by aspirin may prevent the access of arachidonate to the upper part of the active site and thus prevent the cyclooxygenase reaction. Other non-steroidal anti-inflammatory drugs (e.g., ibuprofen, indomethacin, meclofenamate) also compete with arachidonate for binding to the cyclooxygenase active site, perhaps by interacting with the guanidinium group of Arg 120.[27]

Peroxidase Activity

The peroxidase activity of PGH-synthase catalyzes the reduction of the 15-hydroperoxy group of PGG_2, producing PGH_2 (Fig. 4.4). During this reduction, a tyrosine residue (possibly Tyr 385) forms a radical that may participate in activating the cyclooxygenase reaction. This would explain why the presence of low concentrations of hydroperoxides are obligatory for the expression of cyclooxygenase activity.[34,35]

The peroxidase active site in PGH-synthase 1 is highly homologous to other peroxidases (e.g., dog myeloperoxidase) and it is physically distinct from the cyclooxygenase active site.[27,34,35] In agreement with this idea, potent inhibitors of cyclooxygenase activity, such as non-steroidal anti-inflammatory drugs, have no effect on the peroxidase activity of PGH-synthase. The two sites may interact, however, as strongly suggested both by biochemical evidence and by the position of the peroxidase haem group, revealed by X-ray crystallography. This group is located within the peroxidase active site but lies at a distance of only about 10 Å from the γ carbon of Tyr 385.[27]

Pharmacological Inhibition of PGH-Synthase

The non-steroidal drugs that are currently used in the treatment of acute and chronic inflammatory disorders exert their therapeutic effects by inhibiting the cyclooxygenase activity of both PGH-synthase 1 and PGH-synthase 2.[36,37] Cyclooxygenase inhibition results in decreased production of the prostanoids, whose biological actions contribute to the leukocyte infiltration, edema and

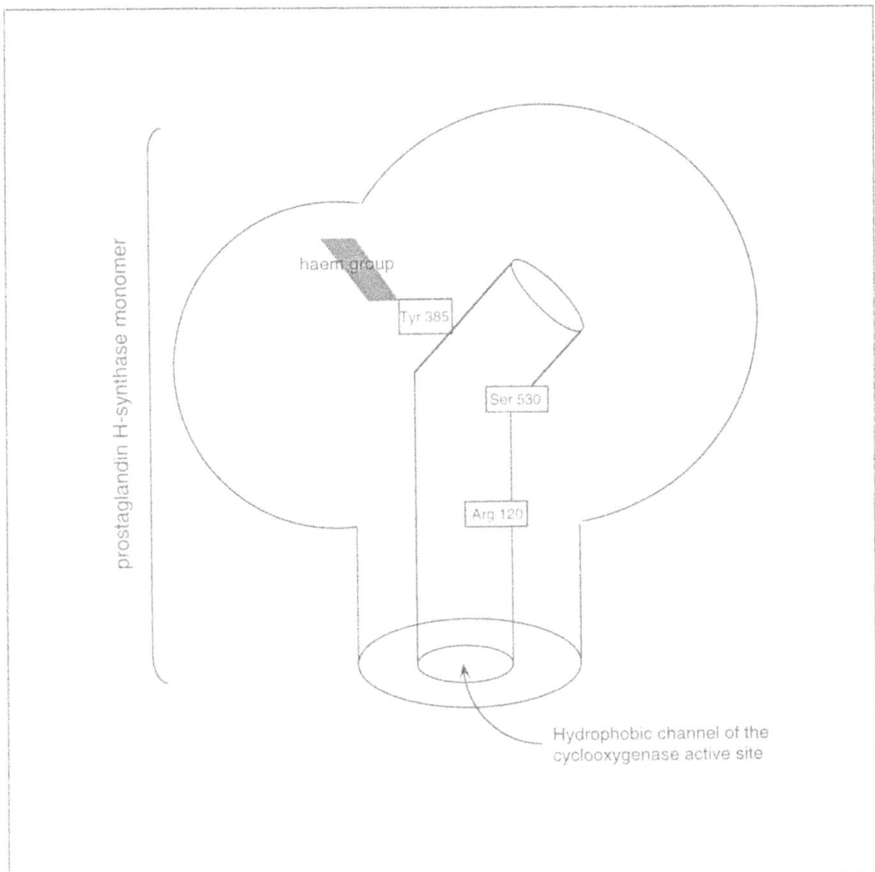

Fig. 4.5. Schematic view of the cyclooxygenase active site of PGH-synthase. For clarity, only one of the two PGH-synthase monomers constituting the enzyme is shown in figure. Based on X-ray crystallographic images published in ref. 27.

pain associated with inflammatory responses. While beneficial at inflammatory sites, decreased biosynthesis of the prostanoids in stomach and kidney, where these compounds play important homeostatic roles, can produce gastric lesions, bleeding and nephrotoxicity. How may one circumvent these debilitating side effects? One possibility is by using drugs that selectively inhibit

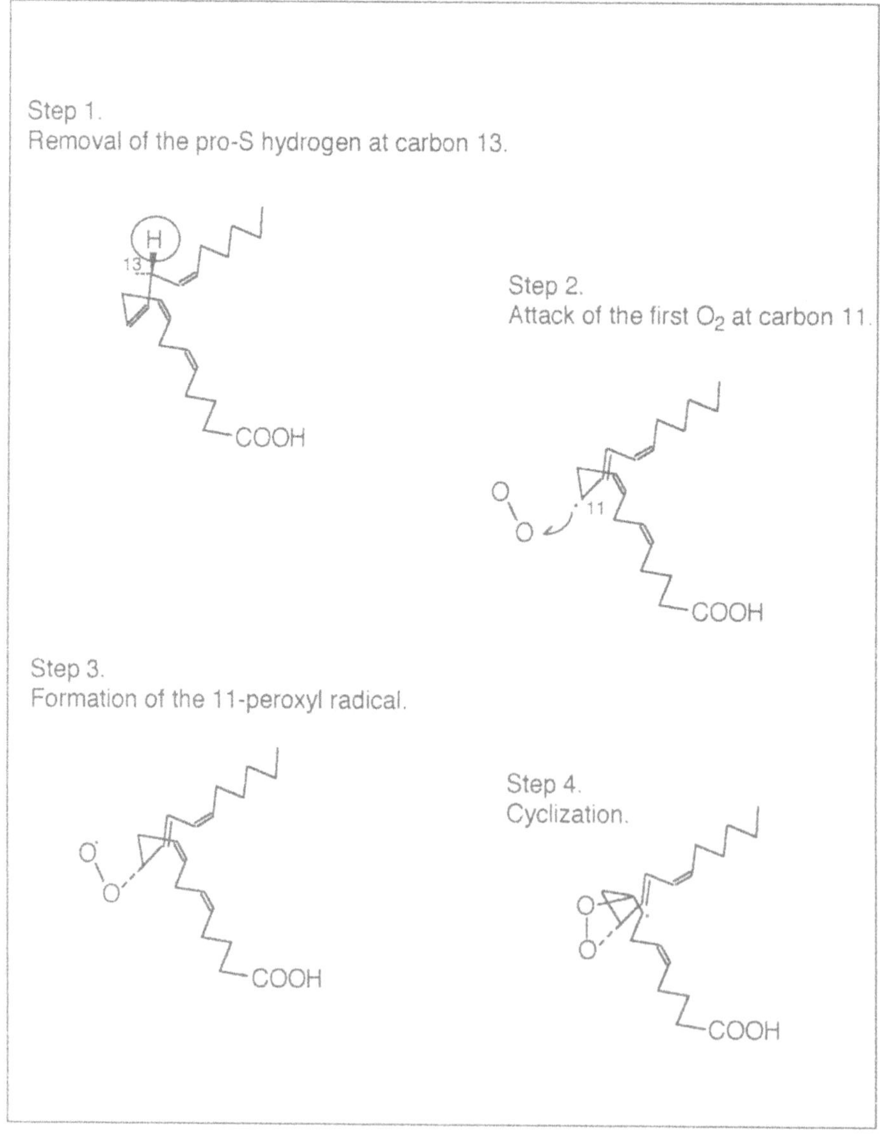

Fig. 4.6a. Proposed mechanism of the cyclooxygenase reaction. See text for details.

the cyclooxygenase activity of PGH-synthase 2 over that of PGH-synthase 1.

We have pointed out that PGH-synthase 1 is a constitutive enzyme, and it is thought to participate in physiological processes such as water and sodium reabsorption in the kidney, gastroprotection in the stomach and vascular homeostasis.[38,39] By contrast, the expression of PGH-synthase 2 is normally very low, but it can be induced by mitogenic and proinflammatory stimuli. For example, in models of acute inflammation, the levels of PGH-synthase 2 mRNA and protein are dramatically enhanced in certain cell types (e.g., macrophages) and so is prostanoid biosynthesis. These results have led to the hypothesis, illustrated in Figure 4.7, that the prostanoids produced by PGH-synthase 2 are those most directly involved in mediating inflammatory responses. This hypothesis has found support in experiments with selective PGH-synthase 2 inhibitors, which were found to be potent anti-inflammatory agents devoid of ulcerogenic properties.[40,41] However, the genetic disruption of PGH-synthase 1 or PGH-synthase 2 by

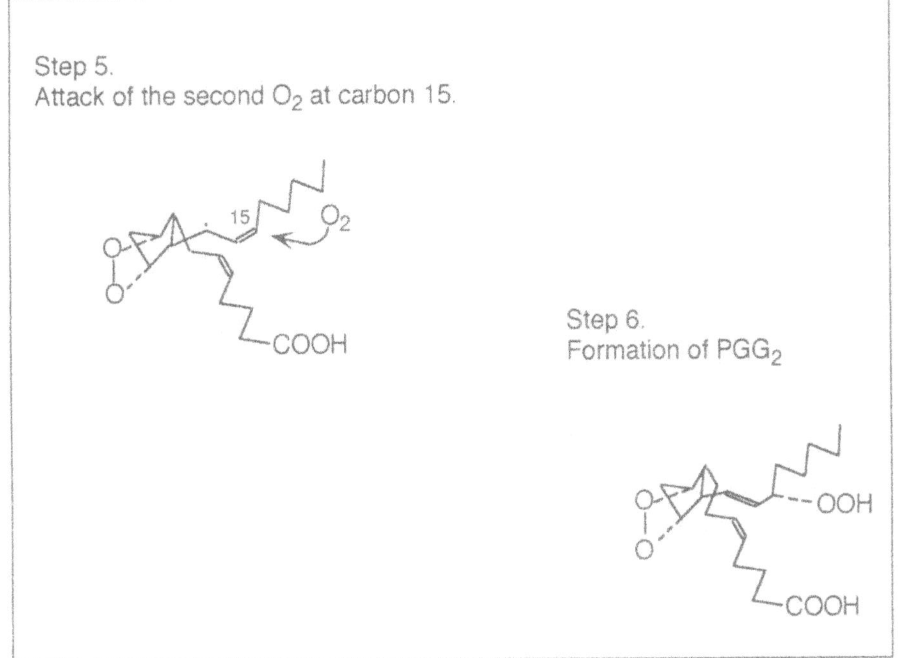

Fig. 4.6.b. Proposed mechanism of the cyclooxygenase reaction.

homologous recombination, recently reported, may lead to reconsider this hypothesis critically.[42,43] The distinct phenotypes of the two mouse strains obtained in these knockout experiments confirm the crucial physiopathological roles of the prostanoids, but do not support the model depicted in Figure 4.7. Mice lacking PGH-synthase 1 have few phenotypic abnormalities but show decreased responses in acute models of inflammation.[42] On the other hand, mice lacking PGH-synthase 2 display normal inflammatory responses but die very young of severe renal failure.[43] These results are exactly opposite to those one would have expected from previous pharmacological data and will remain difficult to interpret till a greater variety of inflammatory insults are tested on these recombinant mice.[44]

Prostaglandin D (PGD)-Synthase

PGD-synthases (also known as PGH/D isomerases) are a group of enzymes that catalyze the conversion of PGH_2 to PGD_2, a reaction that can also occur non-enzymatically under appropriate conditions (Fig. 4.8).

Two distinct types of PGD-synthases have been identified. Y. Urade, T. Shimizu and O. Hayaishi have identified a glutathione-independent PGD-synthase in rat brain tissue, purified it and cloned its complementary DNA.[45-47] This DNA encodes for a membrane-bound polypeptide that is mainly expressed in brain oligodendrocytes.[47] Brain PGD-synthase has a high degree of sequence homology with the lipocalins—a superfamily of extracellular lipid-binding proteins briefly discussed in chapter 3. This sequence similarity can be as high as 50%, suggesting that PGD-synthase may have evolved as a member of the lipocalin superfamily, and may have diverged from it by acquiring enzyme properties and membrane-association.[48] The possible physiological roles of brain PGD-synthase have been thoroughly investigated in Osamu Hayaishi's laboratory, and an involvement of PGD_2 in regulating the levels of cyclic AMP and in the sleep/wake cycle, through selective membrane receptors, has been suggested.[49,50]

A second PGD-synthase, dependent on glutathione for its activity, has been purified from rat spleen in the laboratory of D.H. Nugteren in Holland. The enzyme has a pH optimum between

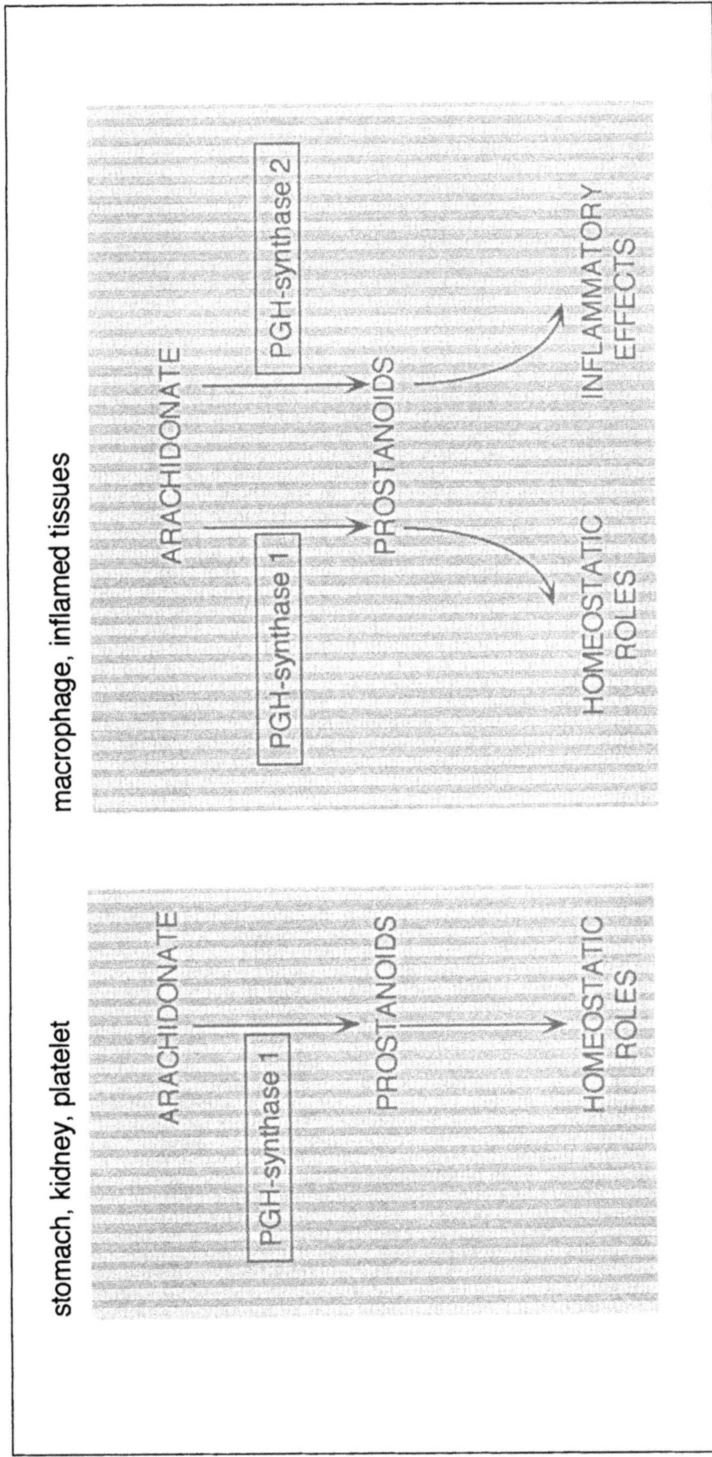

Fig. 4.7. Hypothetical model for the regulation of prostanoid biosynthesis in normal and inflammatory states. Based on results published in ref. 40.

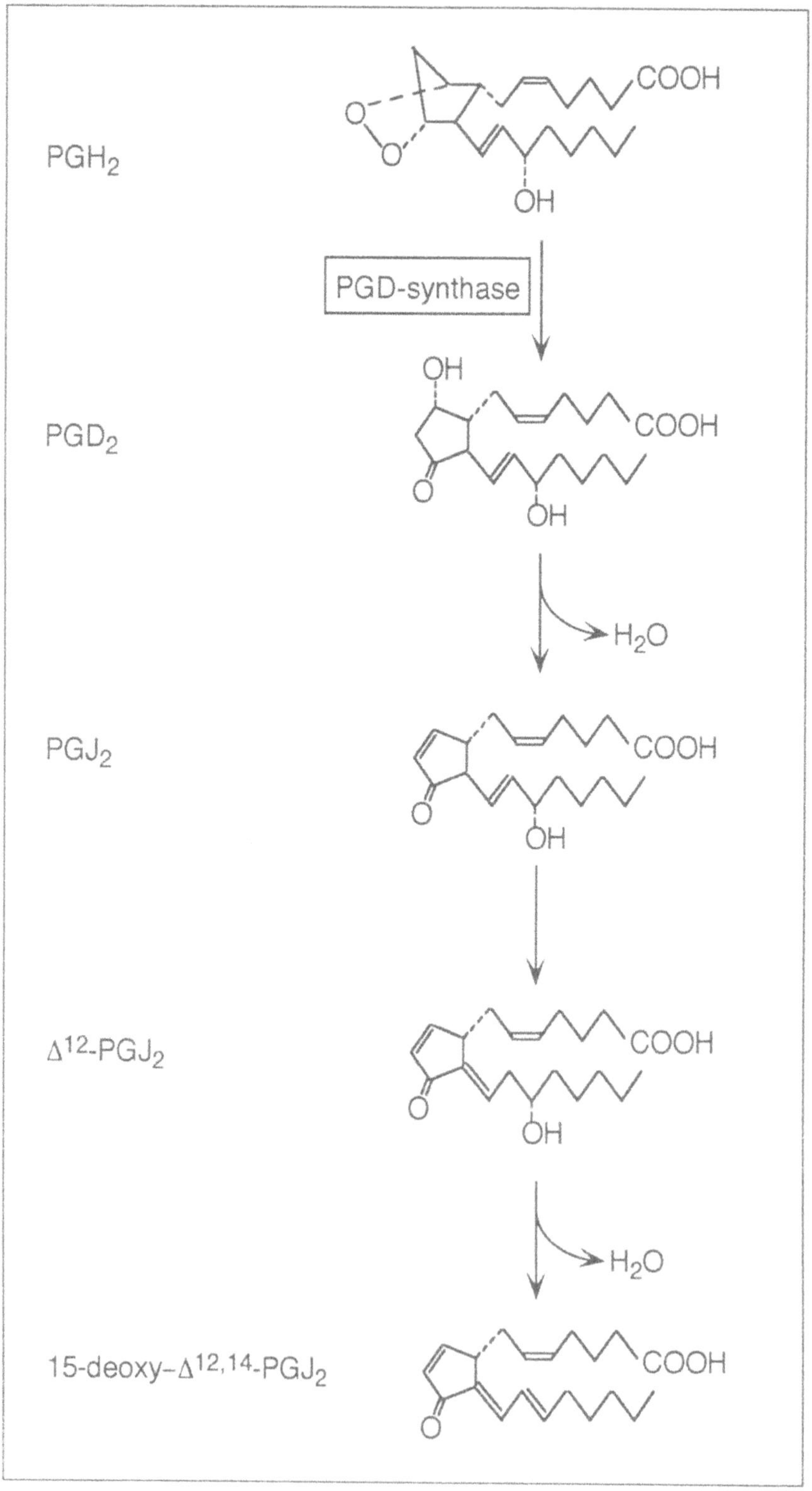

PGH_2

COOH

OH

PGD-synthase

PGD_2

OH

COOH

O

OH

H_2O

PGJ_2

COOH

O

OH

Δ^{12}-PGJ_2

COOH

O

OH

H_2O

15-deoxy-$\Delta^{12,14}$-PGJ_2

COOH

O

Fig. 4.8. Biosynthesis of PGD_2 and its metabolism to J-series prostaglandins.

7 and 8 and is sensitive to cysteine-alkylating reagents.[51] Glutathione-dependent PGD-synthase is thought to be responsible for the formation of PGD_2 in mast cells, and may participate in the mechanisms of inflammation and cutaneous hyperalgesia.[52] In addition to these specific PGD-synthases, the conversion of PGH_2 into PGD_2 may be catalyzed, non-selectively, by glutathione-S-transferases. These enzymes can also isomerize PGH_2 to PGE_2 and $PGF_{2\alpha}$.[53]

The metabolism of PGD_2 to prostaglandins of the J series is shown in Figure 4.8. Three subsequent steps, two of dehydration and one of double bond migration, lead from PGD_2 to 15-deoxy-$\Delta^{12,14}$-PGJ_2. This biochemical pathway may be of considerable physiological significance, as 15-deoxy-$\Delta^{12,14}$-PGJ_2 is considered to be the long-sought endogenous ligand for a nuclear receptor involved in cell differentiation, the peroxisome proliferator-activated receptor.[54-56] We will come back to this subject in the next chapter.

PGD_2 may undergo two additional reactions, which are illustrated in Figure 4.9. It may be converted into $PGF_{2\alpha}$ by the enzyme PGD 11-ketoreductase, a cytosolic protein that requires reduced pyridine nucleotides (NADPH) for activity. The high levels of PGD 11-ketoreductase in liver tissue suggests that the main role

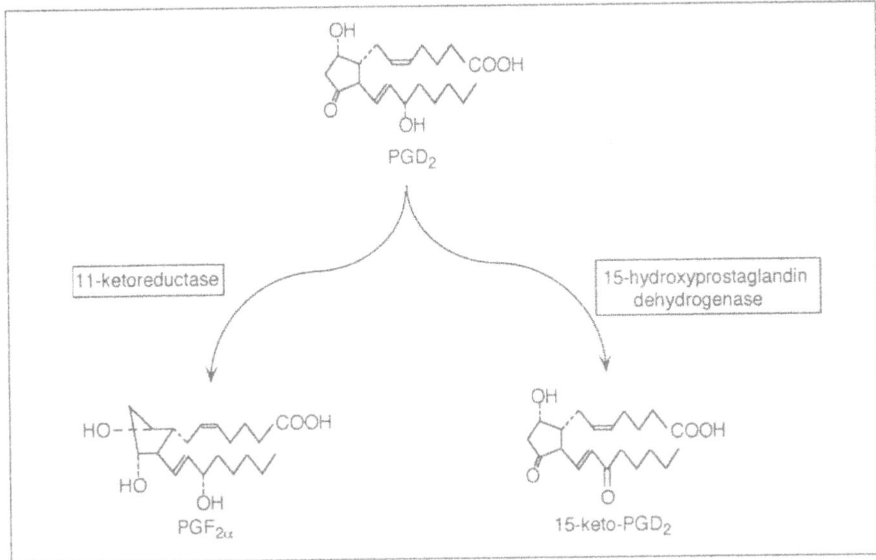

Fig. 4.9. Metabolism of PGD_2 catalyzed by the enzymes PGD_2 11-ketoreductase and 15-hydroxyprostaglandin dehydrogenase.

of this enzyme is catabolic.[57] A PGD 11-ketoreductase activity is also present in brain.[58] Alternatively, PGD_2 may be converted into 15-keto-PGD_2 by a specific 15-hydroxyprostaglandin dehydrogenase, a cytosolic enzyme that requires $NADP^+$(Fig. 4.9).[59]

Prostaglandin E (PGE)-Synthase

The biosynthesis of PGE_2, the major prostaglandin produced in many cell types, is catalyzed by the enzyme PGE-synthase. This reaction involves a non-oxidative rearrangement of the 9,11-peroxido group of PGH_2, which can also occur spontaneously (Fig. 4.10). In vivo, however, most PGE_2 is thought to be formed enzymatically.

PGE-synthase is an integral membrane protein present at high concentrations in microsomal preparations obtained from the vesicular gland of various mammalian species. It has a pH optimum of 5.5-7, and its activity is stimulated by reduced glutathione.[60,61] It may exist in two kinetically and immunologically distinct isoforms.[62] Little is known on the molecular structure of PGE-synthase, though PGE_2 is considered an important autacoid involved, as we have noted above, in the regulation of vascular tone, renal function, inflammation and immunity.

The metabolism of PGE_2 is shown in Figure 4.10. It begins with the oxidation of the 15-hydroxyl group of PGE_2, which can be catalyzed by two distinct isoforms of the enzyme 15-hydroxyprostaglandin dehydrogenase. Type I 15-hydroxyprostaglandin dehydrogenase is a soluble protein.[63-66] Its low ($\approx 1\ \mu M$) K_M for prostaglandins of the E-series and its abundance in tissues that produce high levels of PGE_2 (stomach, lung and kidney) indicate that type I 15-hydroxyprostaglandin dehydrogenase is a major enzymatic system responsible for the catabolism of PGE_2. By contrast, the type II isoform has a very high K_M for E-series prostaglandins ($\approx 500\ \mu M$), and its exact physiological role, if any, remains uncertain.[67] 15-keto-PGE_2 is metabolized further, by reduction and β-oxidation, before being excreted in urine (Fig. 4.10).[68] In alternative to the 15-hydroxydehydroxygenase route, PGE_2 may be metabolized by an NADPH-dependent monooxygenase to yield 19- and 20-hydroxy-PGE_2.[69]

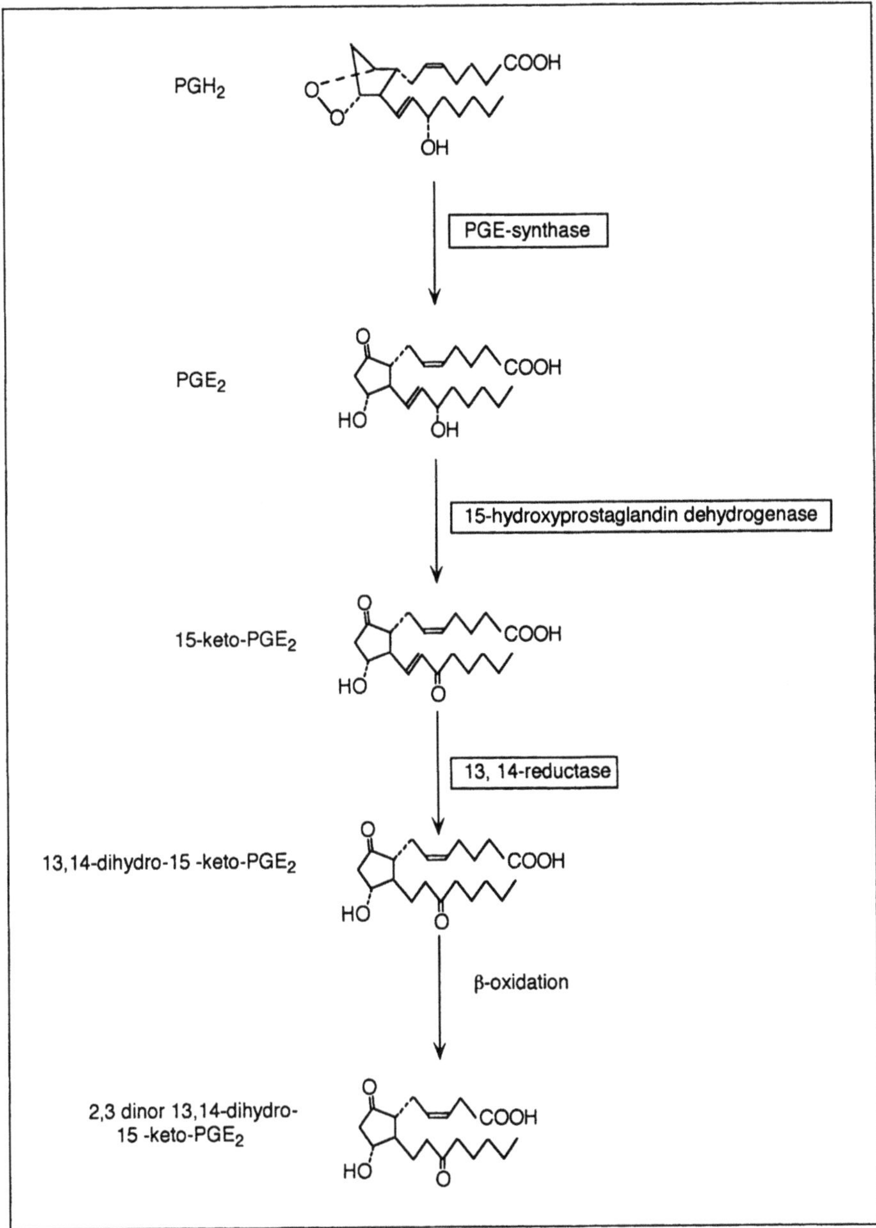

Fig. 4.10. Formation and metabolism of PGE_2.

Prostacyclin-Synthase

Prostacyclin-synthase catalyzes the conversion of PGH_2 to prostacyclin (PGI_2), which is hydrolyzed rapidly and non-enzymatically to the stable product, 6-keto-$PGF_{1\alpha}$[70,71] (Fig. 4.11). Prostacyclin-synthase is a membrane-bound hemoprotein localized in microsomes, and it belongs to the cytochrome P_{450} family.[72,73] Expression of prostacyclin-synthase is particularly high in cells of the vascular endothelium, which synthesize PGI_2 very actively and release it into the blood stream. PGI_2 exerts potent anti-thrombotic and vasodilatory effects by binding to a selective receptor present on the surface of blood platelets, vascular smooth muscle and other cells[72] (see chapter 5).

Like PGE_2 and PGD_2, PGI_2 undergoes multiple metabolic transformations in vivo, after its non-enzymatic conversion to 6-keto-$PGF_{1\alpha}$.[65] Figure 4.11 depicts schematically one such pathway. Through sequential steps of β-oxidation, 15-dehydrogenation and 13,14 reduction, 6-keto-$PGF_{1\alpha}$ is converted into 2,3 dinor-6,15-diketo-13,14-dihydro-$PGF_{1\alpha}$, a metabolite recovered in urine.[68]

Thromboxane A (TXA)-Synthase

The biosynthesis of TXA_2 from PGH_2 is shown in Figure 4.12. Cleavage of the 9,11-peroxido group of PGH_2 is followed by cyclization, which gives rise to the oxane ring structure of TXA_2. TXA-synthase was purified to homogeneity from human platelets, and a complementary DNA encoding it was cloned from human and rat tissues. The human platelet enzyme is a membrane hemoprotein which, like prostacyclin-synthase, belongs to the cytochrome P_{450} family, showing a 35% sequence homology with class III A cytochrome P_{450} enzymes.[73-80]

TXA_2 undergoes a rapid chemical hydrolysis that yields TXB_2. The latter can be metabolized further either by the enzyme 11-hydroxy-thromboxane dehydrogenase, producing 11-dehydro-TXB_2, or by β-oxidation, producing 2,3-dinor-TXB_2 (Fig. 4.12).[68]

TXA_2 is a powerful thrombogenic, vasoconstricting and bronchoconstricting agent produced in relatively large quantities by activated platelets and, to a minor extent, by other tissues (e.g.,

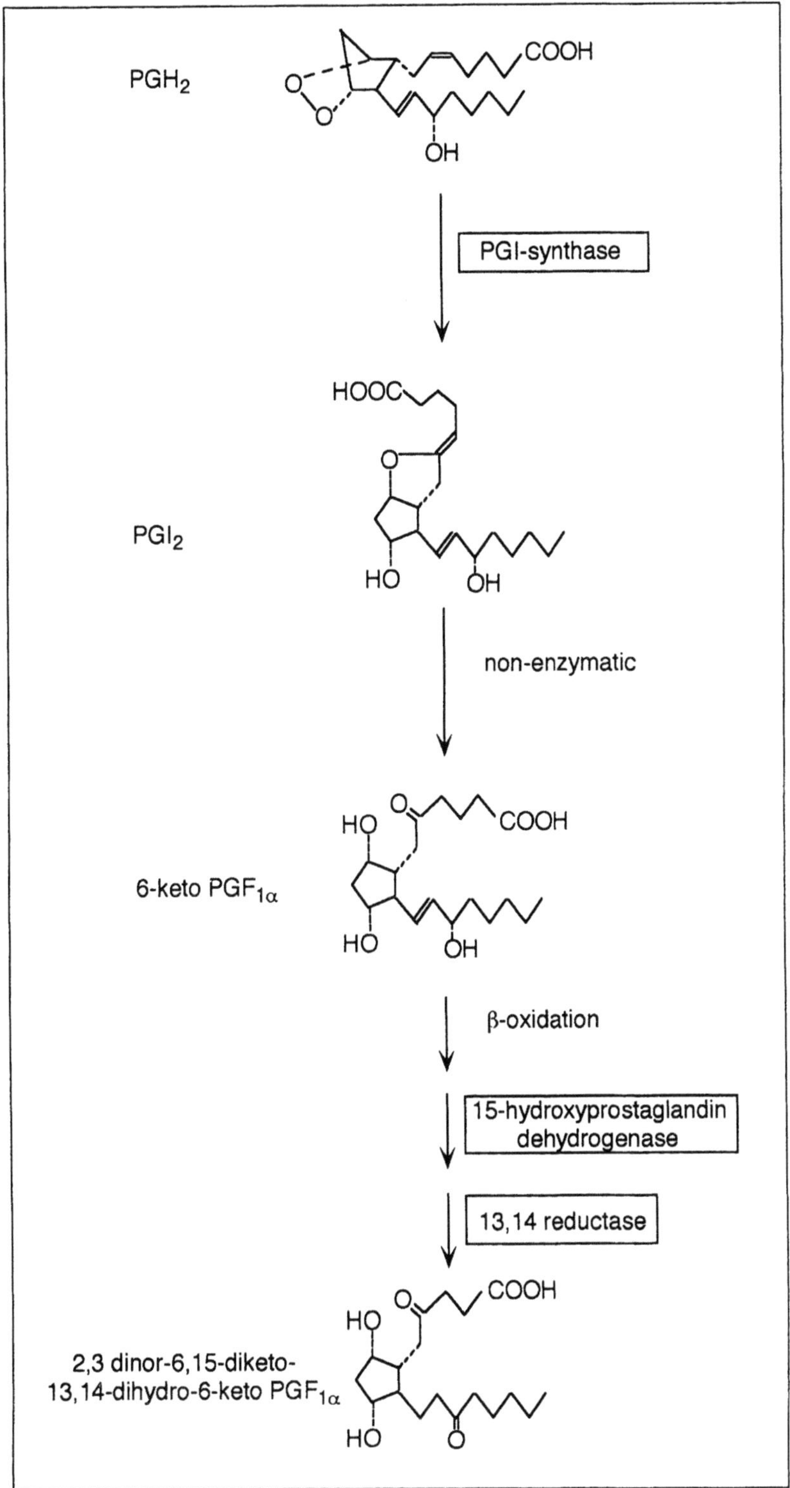

Fig. 4.11. Formation and metabolism of PGI_2.

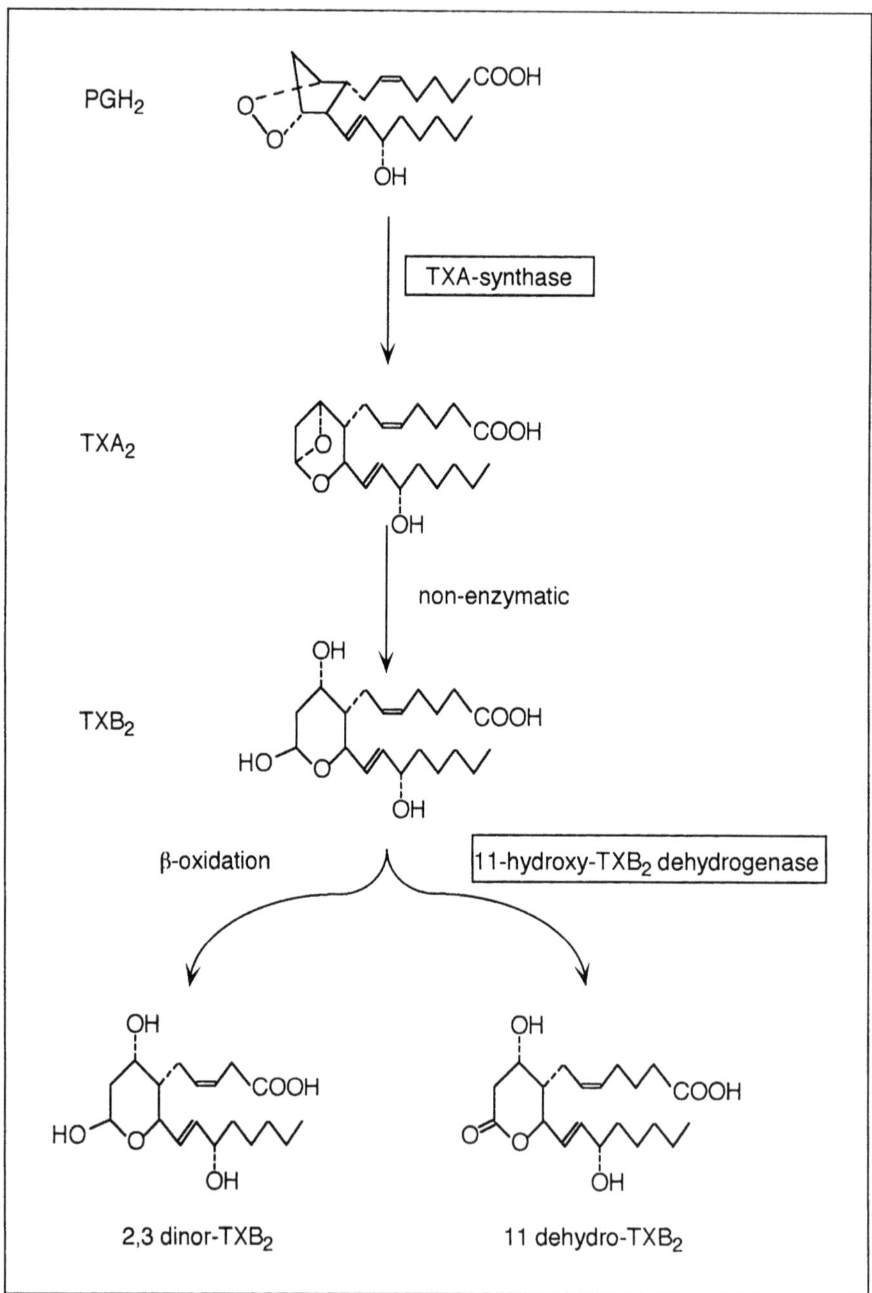

Fig. 4.12. Formation and metabolism of TXA_2.

kidney, intestine and lungs). Because of its potent biological activities, TXA_2 is thought to participate in a variety of pathological conditions, such as stroke and myocardial infarction, and substantial efforts have been made to develop effective TXA-synthase inhibitors. While some of these compounds are now under clinical trial, additional promising drugs may come from combining inhibition of TXA-synthase activity with selective antagonism at the TXA_2 receptor, a G protein coupled membrane receptor that mediates the cellular effects of this eicosanoid (see chapter 5).[81-83]

PGF-SYNTHASE

Two distinct enzymes, PGF-synthase and aldehyde reductase, are thought to participate in the biosynthesis of $PGF_{2\alpha}$. PGF-synthase catalyzes the reduction of either the 9,11-endoperoxide group of PGH_2 or the 9-keto group of PGD_2 (Fig. 4.13). The enzyme, purified and cloned from bovine lung, shows a high degree of homology with human liver aldehyde reductase (39% identity)[84] and, unexpectedly, with the ε-crystalline in frog's lens (58% identity).[85]

Aldehyde reductase converts PGH_2 to $PGF_{2\alpha}$ but, unlike PGF-synthase, it has little or no activity toward PGD_2 (Fig. 4.13). Also, aldehyde reductase does not cross-react with antibodies raised against PGF-synthase, indicating that despite their sequence similarities the two proteins may be enzymatically and structurally different.[86]

THE ISOPROSTANES

The isoprostanes are a family of prostaglandin analogs formed in vivo through the non-enzymatic peroxidation of arachidonate, catalyzed by free-radicals.[87] This reaction may occur primarily in phospholipids, which are then cleaved by phospholipase activities, releasing the free isoprostanes.[88,89] The physiological function, if any, of these compounds is not known, but their production in vivo and the biological actions associated with at least one of them, 8-epi-$PGF_{2\alpha}$ (a vasoconstrictor in kidney, Fig. 4.14),

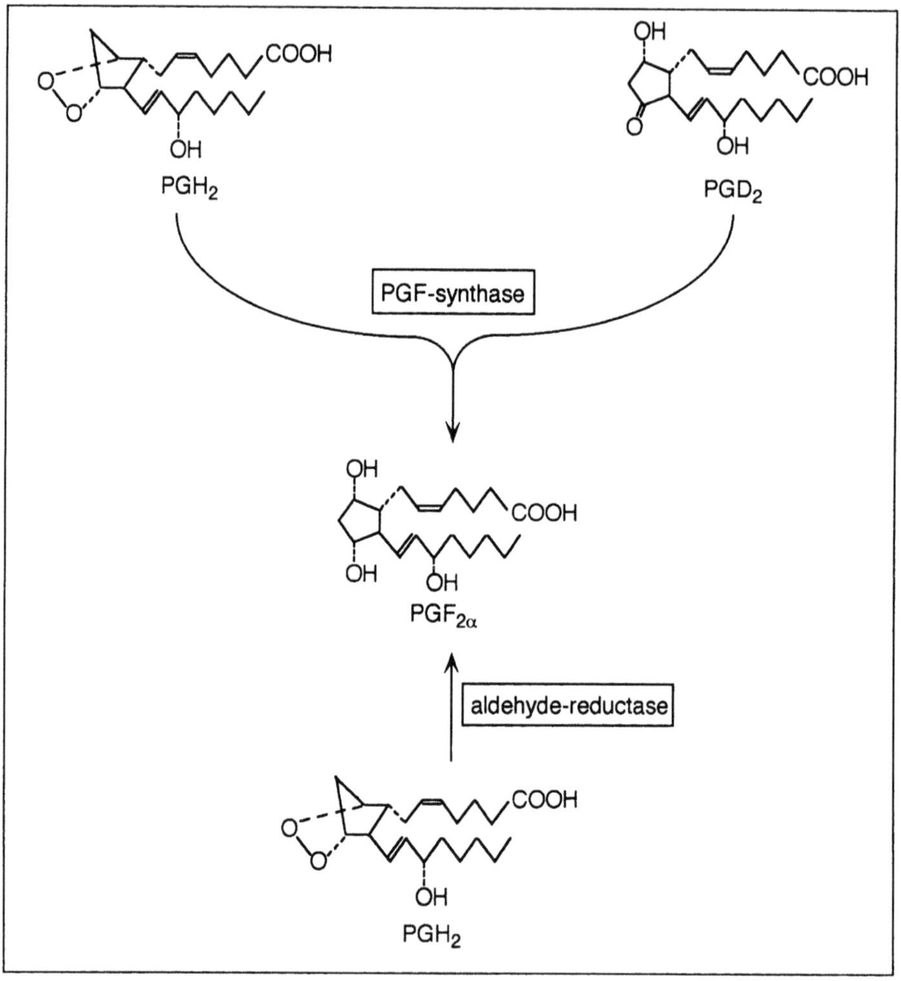

PGH₂ → PGF-synthase ← PGD₂; PGH₂ → aldehyde-reductase → $PGF_{2\alpha}$

Fig. 4.13. The biosynthesis of $PGF_{2\alpha}$ may be catalyzed by either PGF-synthase (top) or aldehyde reductase activities (bottom).

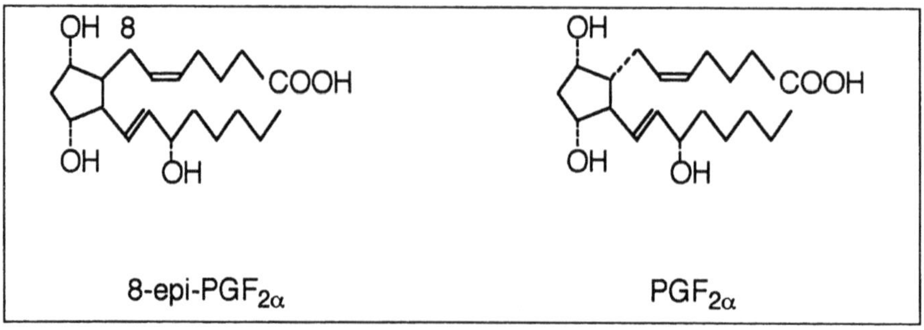

8-epi-$PGF_{2\alpha}$ $PGF_{2\alpha}$

Fig. 4.14. Structure of the biologically active isoprostane, 8-epi-$PGF_{2\alpha}$, and comparison with the structure of the cyclooxygenase-derived product, $PGF_{2\alpha}$.

have led to the suggestion that the isoprostanes may be produced in the course of oxidative insults and play a role in their pathological consequences.[90,91]

THE LIPOXYGENASES

Lipoxygenases catalyze the incorporation of one oxygen molecule into di- or polyunsaturated fatty acids. The reaction occurs at a specific 1,4-*cis,cis*-pentadiene group of a fatty acid substrate, it proceeds *via* the stereoselective elimination of one hydrogen atom followed by the attack of an oxygen molecule and results in the formation of a chiral hydroperoxy-2,4 *trans,cis*-conjugated fatty acid (Fig. 4.15).[92] Lipoxygenase reactions are not only stereoselective, but also positionally specific, and lipoxygenases are usually named after the carbon atom they attack preferentially. Thus, for example, the main product of the arachidonate 12-lipoxygenase is 12(S)-HPETE.[92]

In the next sections, I will recapitulate the properties of a group of three homologous arachidonate lipoxygenases that are thought to serve important physiopathological roles in mammalian tissues. These are the 5-lipoxygenase, the 12-lipoxygenase and the 15-lipoxygenase.

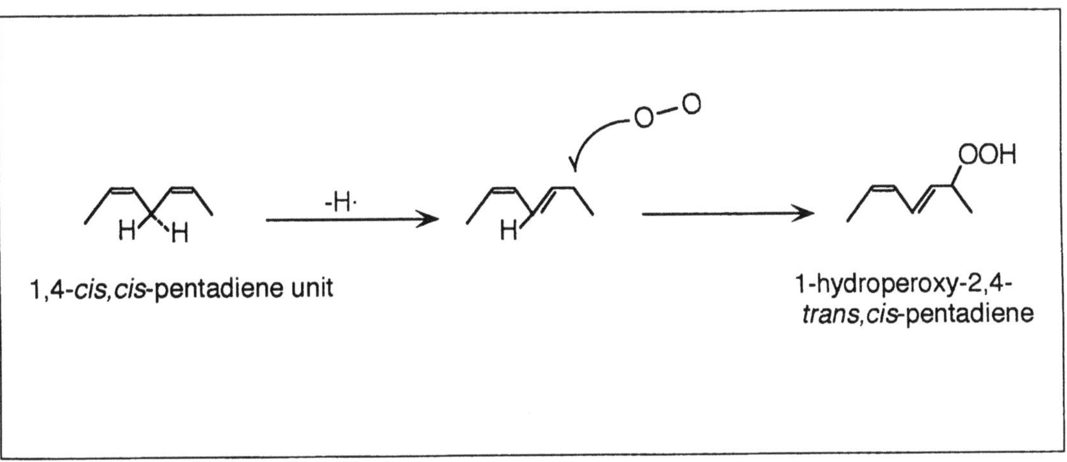

*Fig. 4.15. Proposed mechanism of the lipoxygenase reaction. Lipoxygenases recognize the 1,4-*cis,cis*-pentadiene structure of fatty acid substrates, and catalyze the stereoselective elimination of one hydrogen atom, followed by incorporation of molecular oxygen. This results in the formation of a chiral hydroperoxy fatty acid.*

5-LIPOXYGENASE AND THE LEUKOTRIENES

The biosynthesis of the leukotrienes involves a series of reactions initiated by 5-lipoxygenase, which catalyzes both the oxygenation of arachidonate to 5(S)-HPETE and the subsequent dehydration of this hydroperoxide intermediate to produce the epoxide, 5(S),6(S)-oxido-7,9,11,15(E,E,Z,Z)-eicosatetraenoic

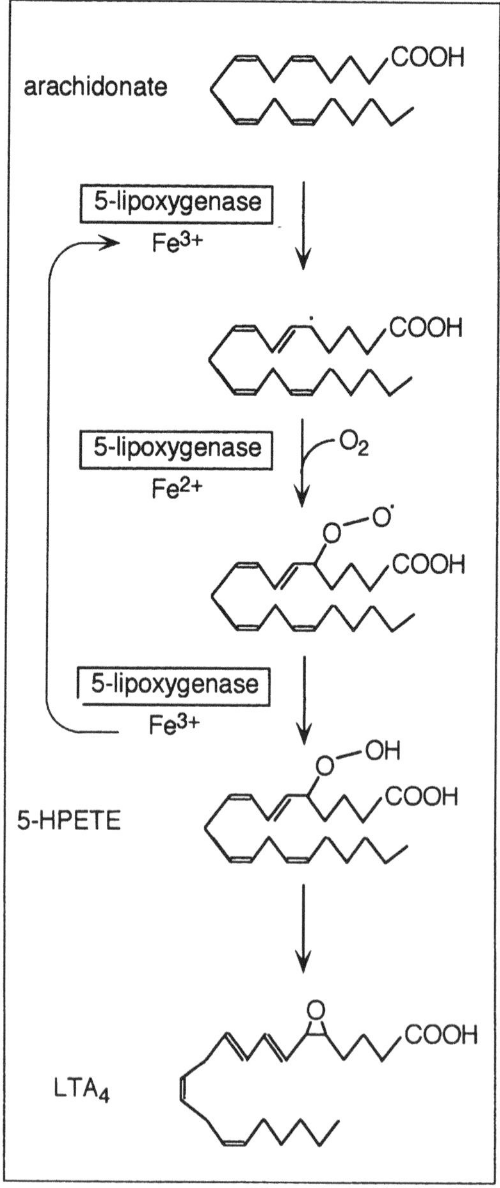

Fig. 4.16. Hypothetical enzyme mechanism of the 5-lipoxygenase reaction. In the resting state, the enzyme contains reduced iron (Fe^{2+}). Enzyme activation occurs when iron is oxidized from Fe^{2+} to Fe^{3+} by hydroperoxides present at the reaction site or formed in situ. The Fe^{3+} enzyme oxidizes arachidonic acid, forming the pentadienyl radical and a proton, while Fe^{3+} is reduced to Fe^{2+}. An oxygen molecule attacks stereoselectively this radical, producing a 5(S)-hydroperoxy radical. Electron transfer and protonation produce 5-hydroperoxyeicosatetraenoic acid (5-HPETE), on the one hand, and Fe^{3+}, on the other. Dehydration of 5-HPETE yields leukotriene A_4 (LTA_4).

acid, better known as leukotriene A_4 (LTA_4) (Fig. 4.16) (for review, see ref. 93). Arachidonate is the natural substrate for 5-lipoxygenase, but other polyunsaturated fatty acids (e.g., 8,11,14 eicosatrienoate) or hydroperoxy fatty acids (e.g., 15-HPETE) may also be used.

As for other lipoxygenases, the first step in the 5-lipoxygenase reaction proceeds through the removal of an hydrogen atom, the pro-S hydrogen at carbon 7. The following step of dehydration also involves stereoselective hydrogen removal (of the pro-R hydrogen at carbon 10) and it is followed by allylic shifts and by donation of an electron and a proton to the leaving hydroxyl group of the hydroperoxyl moiety to produce H_2O and LTA_4 (Fig. 4.16). LTA_4 may be converted to LTB_4, by the enzyme LTA_4 hydrolase, or to LTC_4, by LTC_4 synthase. We will describe each of these steps separately, later in this chapter.

5-Lipoxygenase has been purified to homogeneity, and complementary DNAs encoding it have been cloned and characterized from a number of mammalian tissues and cells. 5-Lipoxygenase is cytosolic, it contains non-heme iron and it is expressed at high levels in cells of myeloid origin (e.g., polymorphonuclear leukocytes, macrophages, etc.).[94-100] Lower levels of expression are seen in other tissues, including select areas of the brain.[101]

The mechanism of activation of 5-lipoxygenase differs substantially from that of other lipoxygenases. Purified 5-lipoxygenase requires both calcium ions and ATP for activity—although the enzyme does not contain any obvious consensus sequence for calcium- or ATP-binding domains. Moreover, maximal 5-lipoxygenase activity is achieved only in the presence of an 18-kDa membrane protein, the 5-lipoxygenase activating protein (FLAP), which was discovered in the laboratory of Anthony W. Ford-Hutchinson at Merck Frosst.

How do these various elements cooperate to produce full 5-lipoxygenase activation? Although a precise answer to this question is not yet possible, a plausible model has been proposed. This model, illustrated in Figure 4.17 and based on evidence reviewed in ref. 93, postulates that binding of calcium and ATP results in the translocation of 5-lipoxygenase from the cytosol to a membrane compartment. There, the enzyme finds its membrane-bound

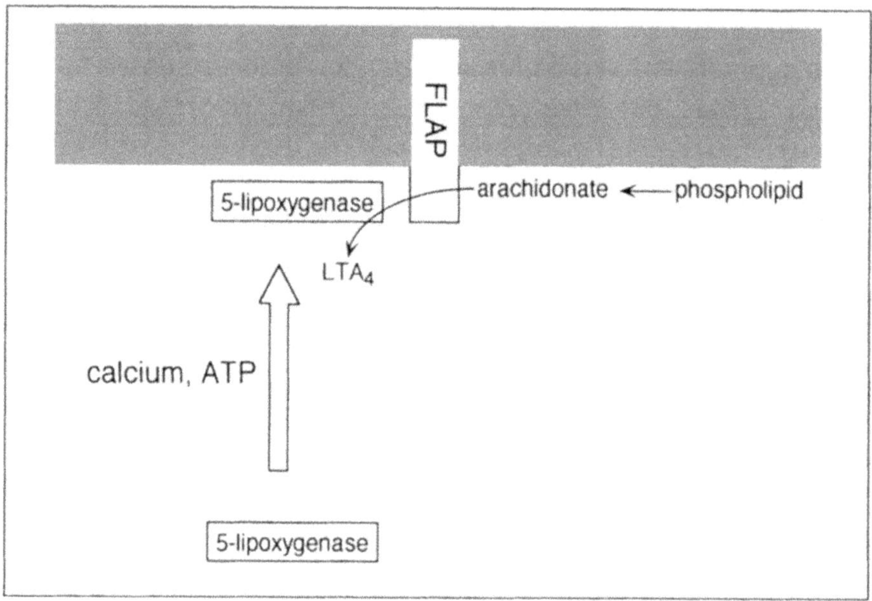

Fig. 4.17. Hypothetical mechanism of 5-lipoxygenase activation by calcium, ATP and 5-lipoxygenase-activating protein (FLAP) in intact cells. Based on results reviewed in ref. 93.

partner, FLAP. FLAP binds to the free arachidonate produced through phospholipase activation and presents it to 5-lipoxygenase, which converts it into LTA_4. After this reaction, the 5-lipoxygenase shuts itself off through a process of self-inactivation.[93]

LTA_4 is the immediate precursor for two distinct groups of biologically active leukotrienes, the peptidoleukotrienes (LTC_4, LTD_4, LTE_4) and the dihydroxyleukotriene, LTB_4. The first committed step in the biosynthesis of the peptidoleukotrienes is catalyzed by LTC_4-synthase.

Leukotriene C_4 (LTC_4)-Synthase

LTC_4-synthase is a unique, membrane-bound glutathione-S-transferase that conjugates LTA_4 with reduced glutathione (Fig. 4.18). Purification to homogeneity and molecular cloning revealed that human LTC_4-synthase is a homodimer composed of two highly hydrophobic 18-kDa subunits. Interestingly, the amino acid sequence of LTC_4-synthase is highly homologous, not to other glutathione-S-transferases as one would have expected, but to FLAP (31% identity, 53% similarity).[102,103] LTC_4 is converted into LTD_4

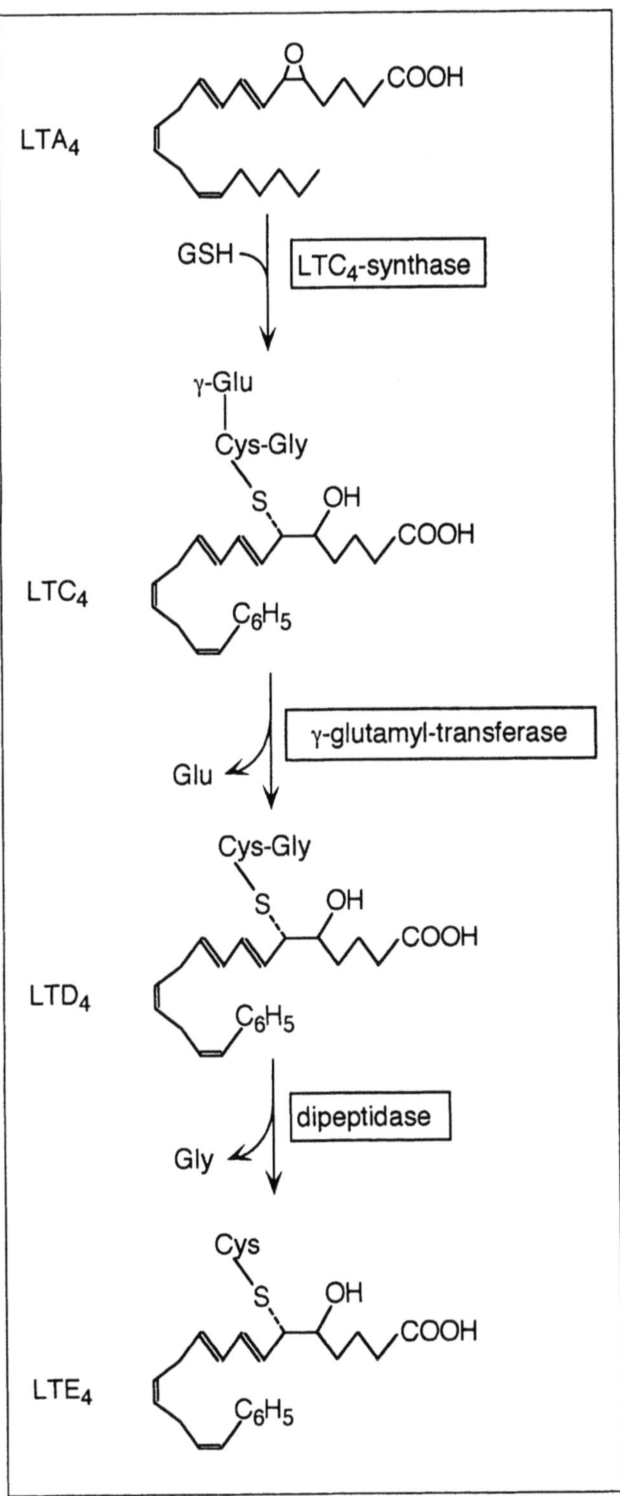

Fig. 4.18. Biosynthesis and metabolism of the peptidoleukotriene, LTC_4.

and LTE_4 by two sequential enzymatic steps catalyzed by γ-glutamyl transferase and LTD_4-dipeptidase, respectively (Fig. 4.18). The latter is a phosphatidylinositol-anchored enzyme and has been purified and cloned from sheep lung tissue.[104]

LEUKOTRIENE A_4 (LTA_4)-HYDROLASE

The hydration of LTA_4 to LTB_4 [5(S),12(R)-dihydroxy-6,14-*cis*-8,10-*trans*-eicosatetraenoic acid] (Fig. 4.19) is catalyzed by

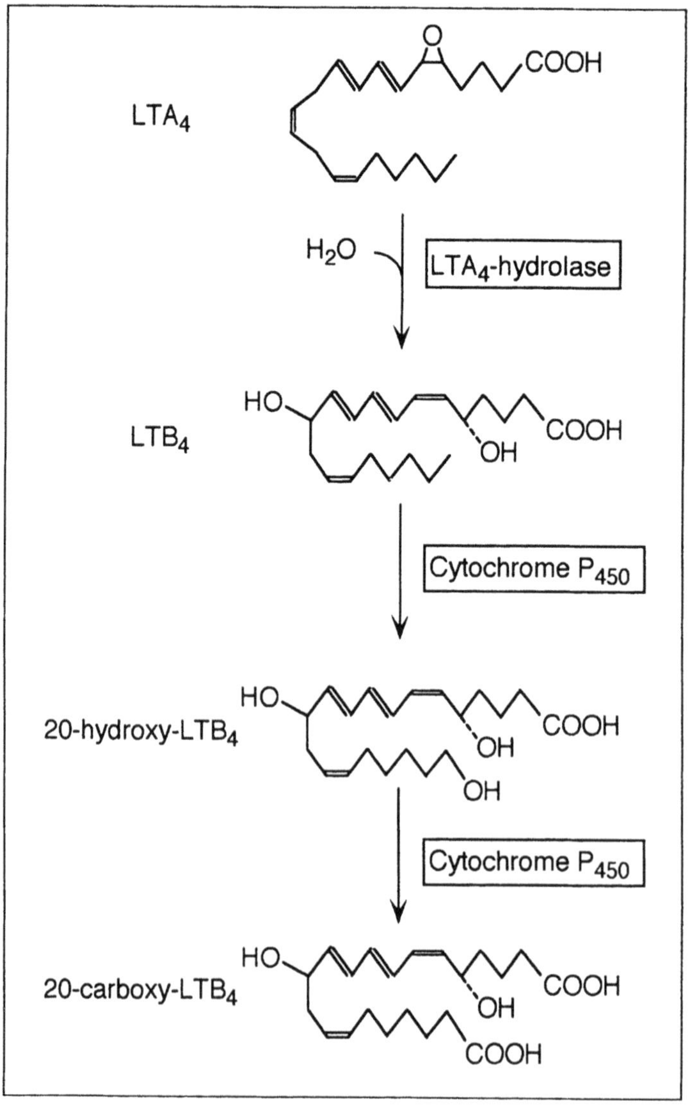

Fig. 4.19. Biosynthesis and metabolism of the hydroxyleukotriene, LTB_4.

LTA_4

H_2O — LTA_4-hydrolase

LTB_4

Cytochrome P_{450}

20-hydroxy-LTB_4

Cytochrome P_{450}

20-carboxy-LTB_4

LTA_4-hydrolase, a specific epoxide hydrolase that has been purified to homogeneity and cloned from a variety of tissues (for review, see ref. 105). LTA_4-hydrolase is a cytosolic enzyme with no apparent cofactor requirement.[105] It is distributed in a number of tissues and cell types, including neutrophils, epithelial cells, smooth muscle cells and enteric neurons,[106] not all of which express 5-lipoxygenase. Cells that lack 5-lipoxygenase, and therefore cannot synthesize LTA_4, can still produce LTB_4, however. They do so via transcellular biosynthesis, i.e., by using LTA_4 released from neighboring blood cells and stabilized by binding to serum albumin.[107]

Two interesting properties of LTA_4-hydrolase, discovered independently in the laboratories of T. Shimizu and F.A. Fitzpatrick, are its sensitivity to inhibitors of zinc metalloproteases (e.g., bestatin) and its ability to catalyze an aminopeptidase reaction.[108,109] Subsequent studies by site-directed mutagenesis have unambiguously demonstrated that LTA_4-hydrolase requires zinc for activity, and have identified its zinc-binding domain.[110] These findings are of considerable pharmacological interest, as they point to the use of known zinc metalloenzyme inhibitors as templates to develop novel blockers of LTB_4 biosynthesis.

Inhibition of Leukotriene Biosynthesis

The leukotrienes are thought to be key elements in common pathological conditions such as inflammation and bronchial asthma. This idea, which stems directly from the fact that the "slow-reacting substance of anaphylaxis" is constituted mainly of leukotrienes (see chapter 1), has found support in a large body of experimental evidence[111]—including genetic disruption of the 5-lipoxygenase gene by homologous recombination.[112] Therefore, drugs that prevent leukotriene formation are considered to have a considerable potential for therapeutic applications.

How can one inhibit leukotriene biosynthesis? From what we have seen above, there are several possible ways (Fig. 4.20). Inhibition of 5-lipoxygenase activity is the most obvious one, and it has been extensively explored. In this line, the design of drugs that interact with the iron present at the active site of 5-lipoxygenase has been particularly fruitful. To this group of compounds belongs zileuton, the 5-lipoxygenase inhibitor that is most advanced in clinical development (as an antiasthmatic agent). Other compounds include 'redox' inhibitors and inhibitors that interfere competitively with the enzyme's active site (for review, see ref. 113). 5-lipoxygenase may also be inhibited indirectly, through the inhibition of its activating protein, FLAP. The prototype of this family of drugs, the compound MK-886, is potent in preventing leukotriene biosynthesis in intact cells whereas, as

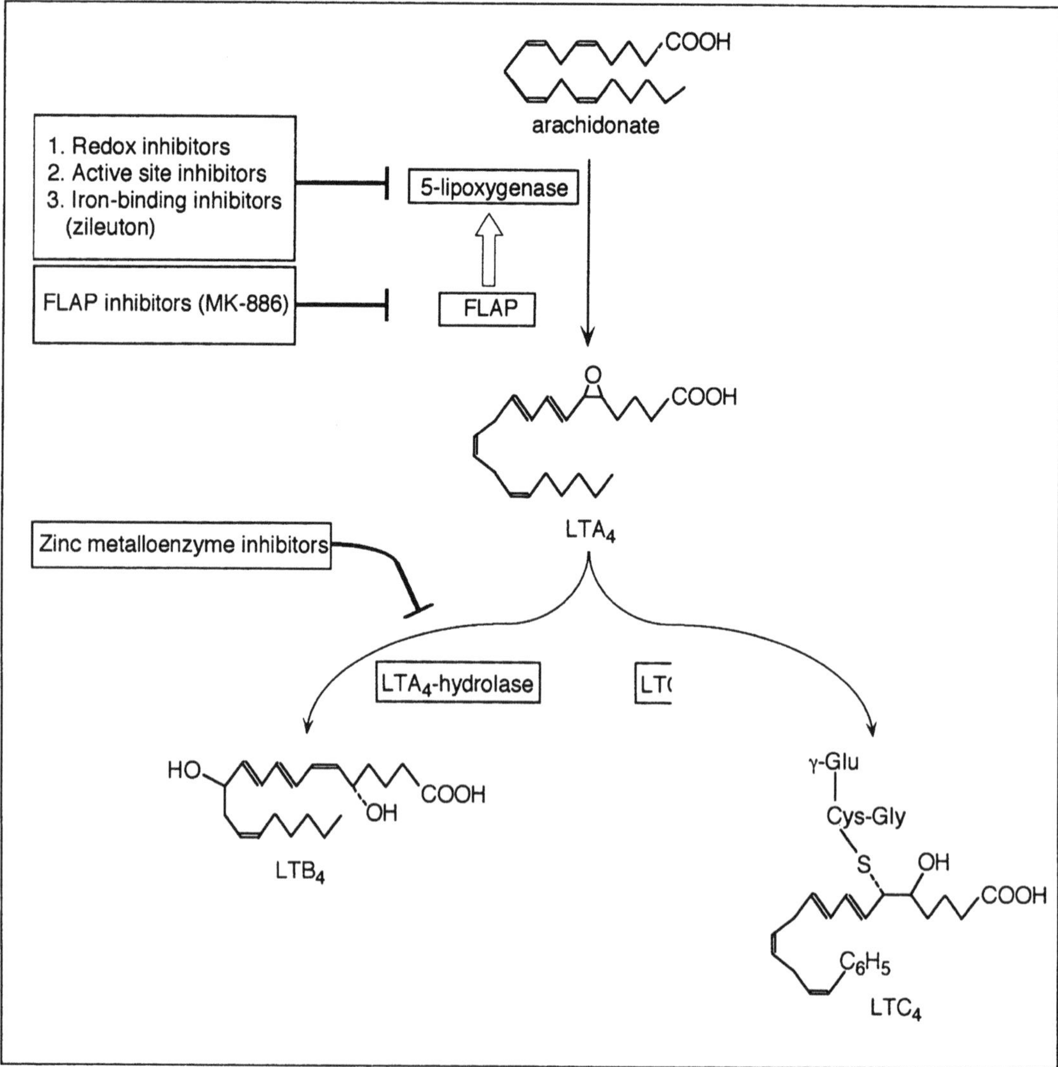

Fig. 4.20. Sites of actions of various leukotriene biosynthesis inhibitors.

expected, it has no effect on the activity of purified 5-lipoxygenase.[114,115] The clinical development of MK-886 has been discontinued because of its toxicity.

In some cases, it may be advantageous to inhibit only one of the branching pathways of LTA_4 metabolism. For instance, in certain inflammatory conditions, LTB_4, not LTC_4, may play a predominant role. In those cases, general inhibition of 5-lipoxygenase activity may be safely replaced by selective inhibition of LTA_4-hydrolase. Taking advantage of the mechanistic similarities between LTA_4 hydrolase and other zinc metalloenzymes, relatively potent inhibitors of this enzyme, which are active both in vitro and in situ, have been now developed.[116]

12-LIPOXYGENASE

12-lipoxygenases are cytosolic proteins that catalyze the conversion of arachidonate to 12(S)-HPETE (Fig. 4.21) although, under certain conditions, 15(S)-HPETE may also be produced as a side product. Three main isoforms—the 'rat platelet type', the 'porcine leukocyte type' and the 'bovine epithelial type'—have been purified, and complementary DNAs encoding them have been cloned.[117-122] The three isoforms have similar molecular masses (≈75 kDa), show a high degree of sequence homology (for example, the platelet and leukocyte types are 65% identical) and have only minor differences in substrate specificity (the platelet type is more specific for arachidonate than the leukocyte type).[123] That these isoforms are not species-specific variants of the same enzyme, as initially suspected, is indicated by several findings, including the expression of the 'porcine leukocyte type' 12-lipoxygenase in rat pineal glands.[124] 12-lipoxygenase activity can be inhibited by reducing compounds, such as nordihydroguaretic acid (NDGA) and baicalein, but these drugs show little selectivity for 12-lipoxygenase over other lipoxygenase activities.

The metabolism of 12(S)-HPETE, illustrated in Figure 4.21, yields two main classes of compounds. The enzymatic reduction of 12(S)-HPETE by glutathione peroxidases gives rise to 12(S)-HETE, a signaling molecule involved in cell proliferation and adhesion.[125-126] Alternatively, structural rearrangements of the hydroperoxy moiety of 12(S)-HPETE produce two distinct hydroxy-epoxides, hepoxilin A_3

(8-hydroxy-11,12-epoxy-5,9,14-eicosatrienoic acid) and hepoxilin B_3 (10-hydroxy-11,12-epoxy-5,8,14-eicosatrienoic acid).[127] Hepoxilin A_3 may act as a second messenger in neural cells and blood platelets, where its biosynthesis is thought to occur through an enzymatic, stimulus-dependent mechanism.[128-134] By contrast, hepoxilin B_3 is produced non-enzymatically, and its biological significance, if any, remains unknown. The hepoxilins are short-lived: an epoxide-hydrolase activity rapidly hydrolyzes their unstable epoxide ring, producing the corresponding trihydroxy acids, trioxilin A_3 and trioxilin B_3 (Fig. 4.21).[127] Additional biotransformations of 12(S)-HPETE

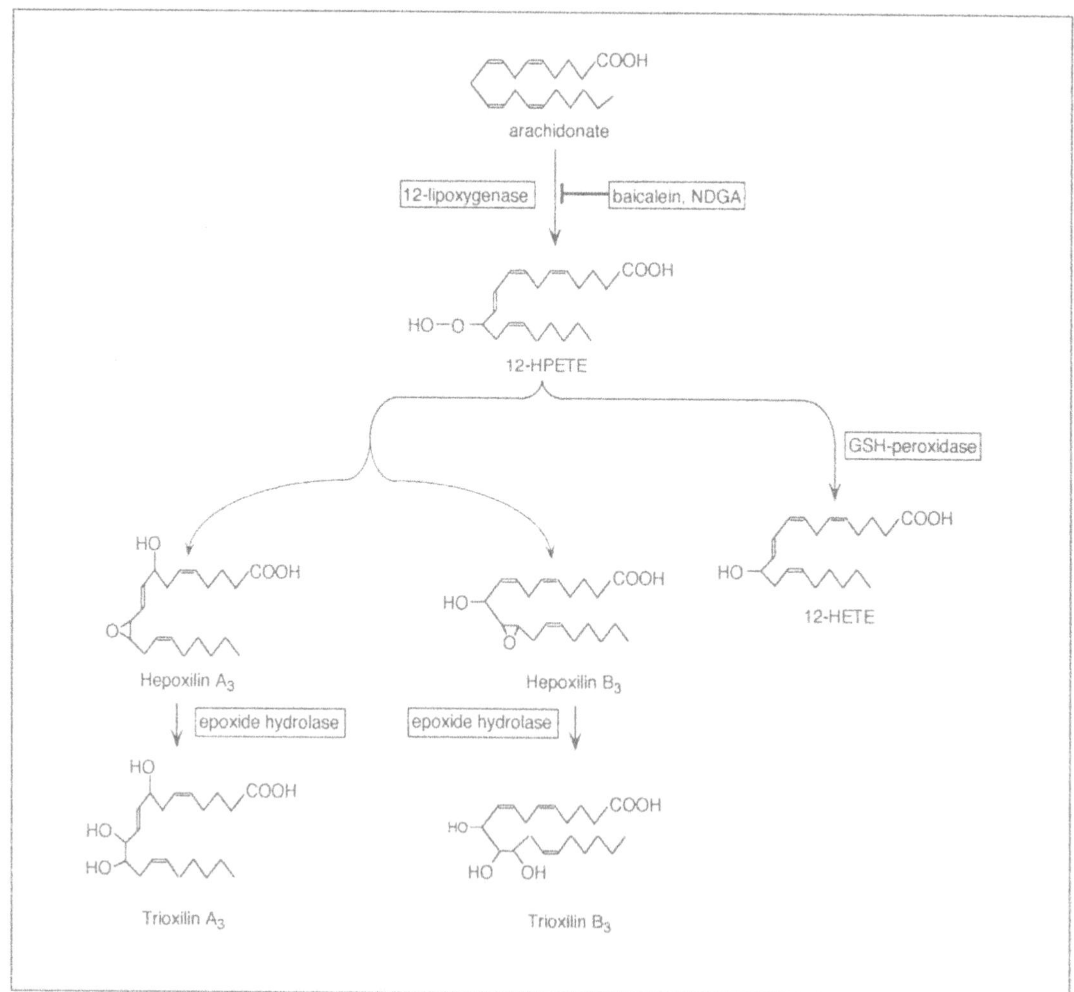

Fig. 4.21. Biosynthesis and metabolism of 12(S)-HPETE.

have been described. For example, in the nervous system of the marine mollusk, *Aplysia californica*, 12(S)-HPETE is converted to the ketoacid, 12-ketoeicosatetraenoic acid (12-KETE), which may participate in cellular signaling.[135]

15-LIPOXYGENASE

15-lipoxygenases catalyze the oxygenation of arachidonate to 15(S)-HPETE, which is further converted, enzymatically or non-enzymatically, to 15(S)-HETE, hydroxy-epoxides and dihydroxy-acids (Fig. 4.22). 15-lipoxygenases have two features that distinguish them from either 5-lipoxygenase or 12-lipoxygenase. First, they are exquisitely active on polyunsaturated fatty acids with 18 carbon atoms, such as linoleic acid (18:2 $\Delta^{9,12}$) (Fig. 4.23). Second, they can oxygenate polyunsaturated fatty acids even when these are esterified to phospholipids or bound to lipoproteins.[136,137]

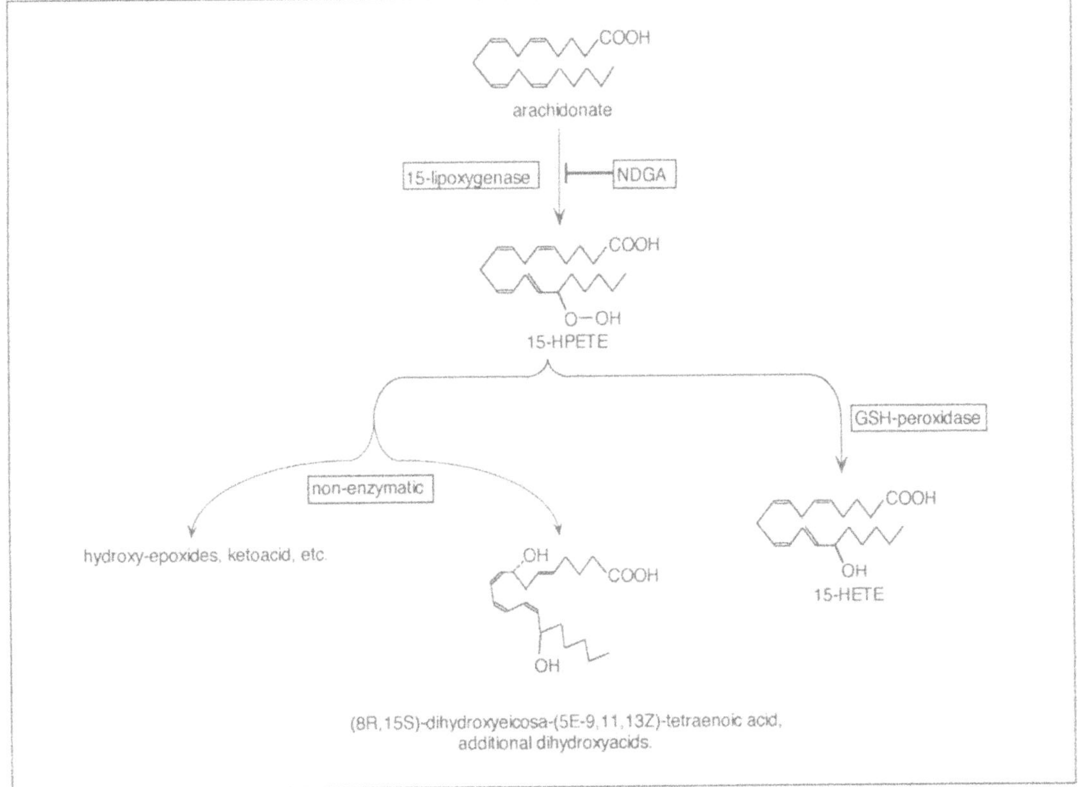

Fig. 4.22. Biosynthesis and metabolism of 15(S)-HPETE.

Different isoforms of 15-lipoxygenase have been purified to homogeneity from mammalian sources, including reticulocytes and leukocytes, and complementary DNAs encoding them have been cloned.[138-142] The reticulocyte 15-lipoxygenase is a cytosolic protein, whose amino acid sequence is 65% similar to human 5-lipoxygenase and 45% similar to type I soybean lipoxygenase (whose tridimensional structure has been resolved by

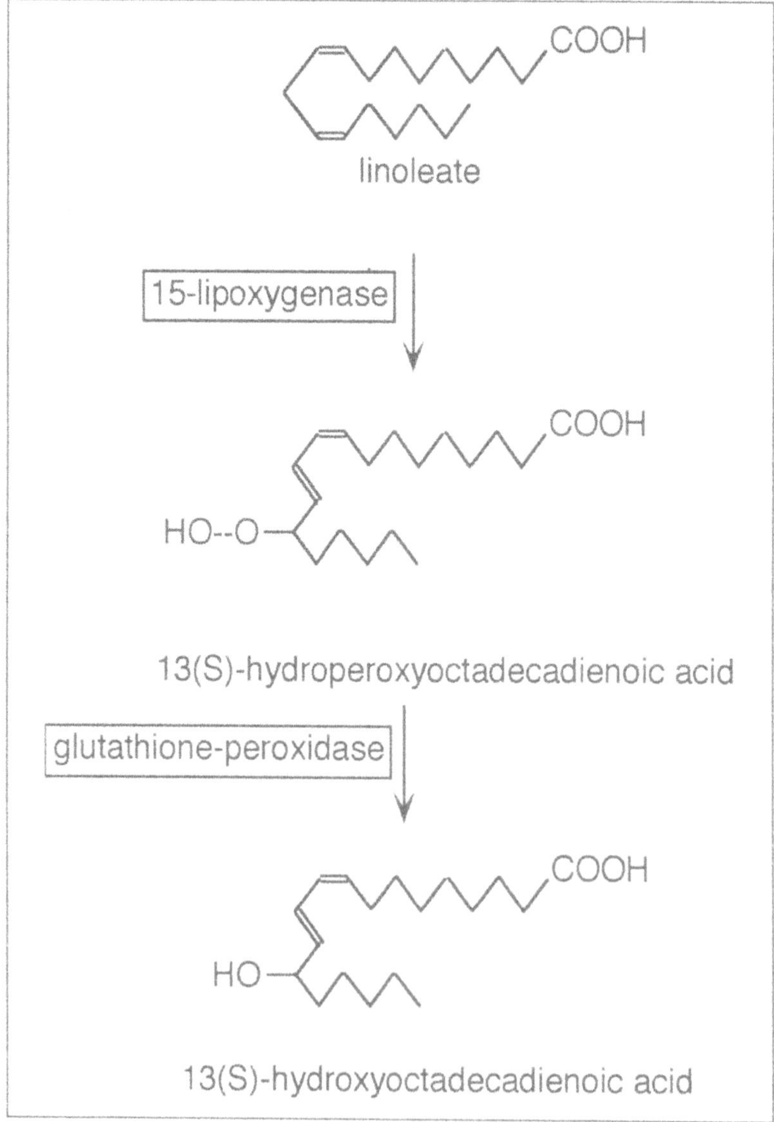

Fig. 4.23. Metabolism of linoleic acid via the 15-lipoxygenase pathway.

X-ray crystallography).[143] Products of arachidonate metabolism via the 15-lipoxygenase pathway have been involved in hyperalgesia and inflammation,[144] while the 15-lipoxygenase-catalyzed oxygenation of esterified fatty acids in phospholipids and in low-density lipoproteins has been implicated in the maturation of reticulocytes to erythrocytes and in the etiology of atherosclerosis (see, for review, ref. 145).

15-Lipoxygenase metabolites of linoleic acid (18:2 $\Delta^{9,12}$) are produced in tissues and may participate in cellular signaling. For example, biosynthesis of 13(S)-hydroxyoctadecadienoic acid (13-HODE) in Syrian hamster embryo fibroblasts is increased when the cells are stimulated with epidermal growth factor (EGF). Moreover, the addition of 13-HODE to the fibroblasts enhances EGF-induced proliferation.[146,147]

Specific 15-lipoxygenase inhibitors are not yet available, but several drugs that inhibit 12-lipoxygenase and 5-lipoxygenase activities (e.g., NDGA) exert a similar inhibitory effect on 15-lipoxygenase.

Products of Multiple Lipoxygenases: The Lipoxins

Before we consider the third main route of arachidonate metabolism in vertebrates, cytochrome P_{450}, we need to describe one last group of lipoxygenase products, the lipoxins (Fig. 4.24).[11] These eicosanoids are unique in that they are produced through a series of enzymatic steps that are catalyzed by different lipoxygenase activities, localized in different cells.

Let us illustrate this rather complex set of interactions with an example. Endothelial cells that line the blood vessel walls produce, upon stimulation, relatively high quantities of the 15-lipoxygenase metabolites, 15(S)-HPETE and 15(S)-HETE. When polymorphonuclear leukocytes adhere to the endothelium, as it happens for example during inflammatory events, extracellularly released 15(S)-HPETE and 15(S)-HETE may enter the leukocytes and serve as substrates for 5-lipoxygenase activity (Fig. 4.25). The product of this double transformation, an unstable 5,6-epoxytetraene, is then converted by epoxide hydrolases to three products, lipoxin A_4, lipoxin B_4 and 7-*cis*,11-*trans*-lipoxin A_4 (Fig. 4.25). Analogous transcellular interactions are thought to occur between other cell types, for example, leukocytes and epithelial cells or leukocytes and platelets.

The lipoxins have a variety of pharmacological effects on the vasculature and the immune system, possibly mediated by selective G protein-coupled membrane receptors (see ref. 148). It is still difficult, though, to point to a specific physiological role for these eicosanoids. One suggestion is that they may act as immunoregulators, inhibiting proinflammatory events while turning on the process of wound healing.[148] The development of selective lipoxin receptor antagonists should provide the tools to address this interesting hypothesis.

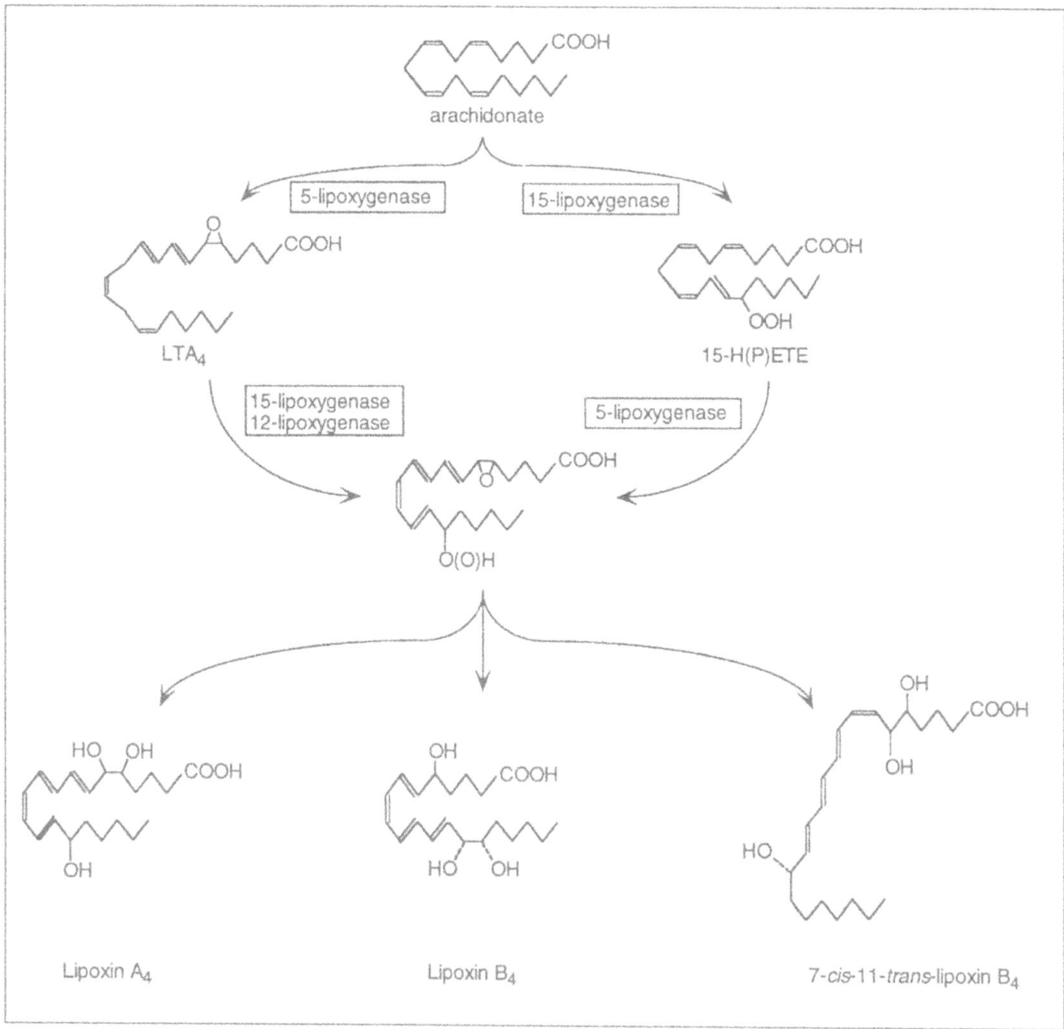

Fig. 4.24. Enzymatic pathways in lipoxin biosynthesis.

CYTOCHROME P_{450}

Cytochrome P_{450} metabolism of arachidonate requires molecular oxygen and NADPH, and consists of two main monooxygenase reactions, illustrated in Figure 4.26 (see ref. 149 for review). First, the epoxidation of the arachidonate double bonds results in the formation of four isomeric epoxyeicosatrienoic acids: 5,6-, 8,9-, 11,12-, and 14,15-EET. These metabolites are subsequently hydrolyzed by an epoxide hydrolase activity to yield

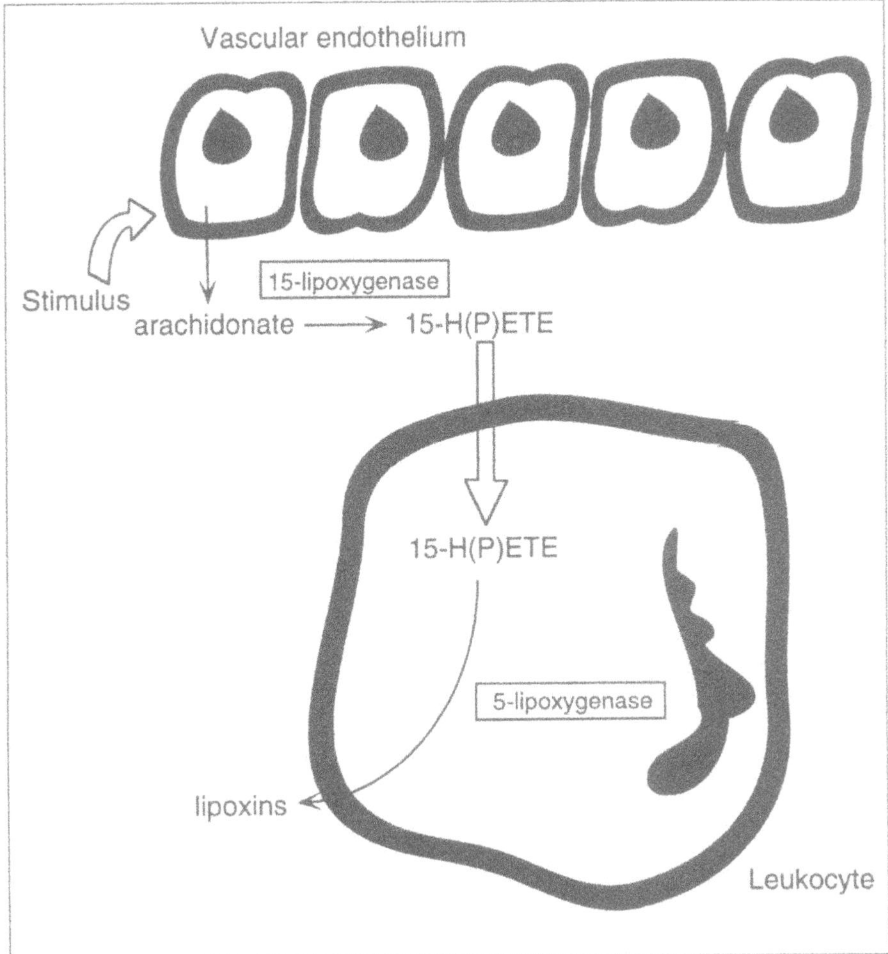

Fig. 4.25. Transcellular biosynthesis of the lipoxins: a hypothetical model. The 15-lipoxygenase metabolites released from endothelial cells, 15(S)-HPETE and 15(S)-HETE, are metabolized by leukocyte 5-lipoxygenase to produce the lipoxins. These, in turn, may act by modulating cell-cell interactions during inflammatory and immune responses. Based on data reviewed in ref. 148.

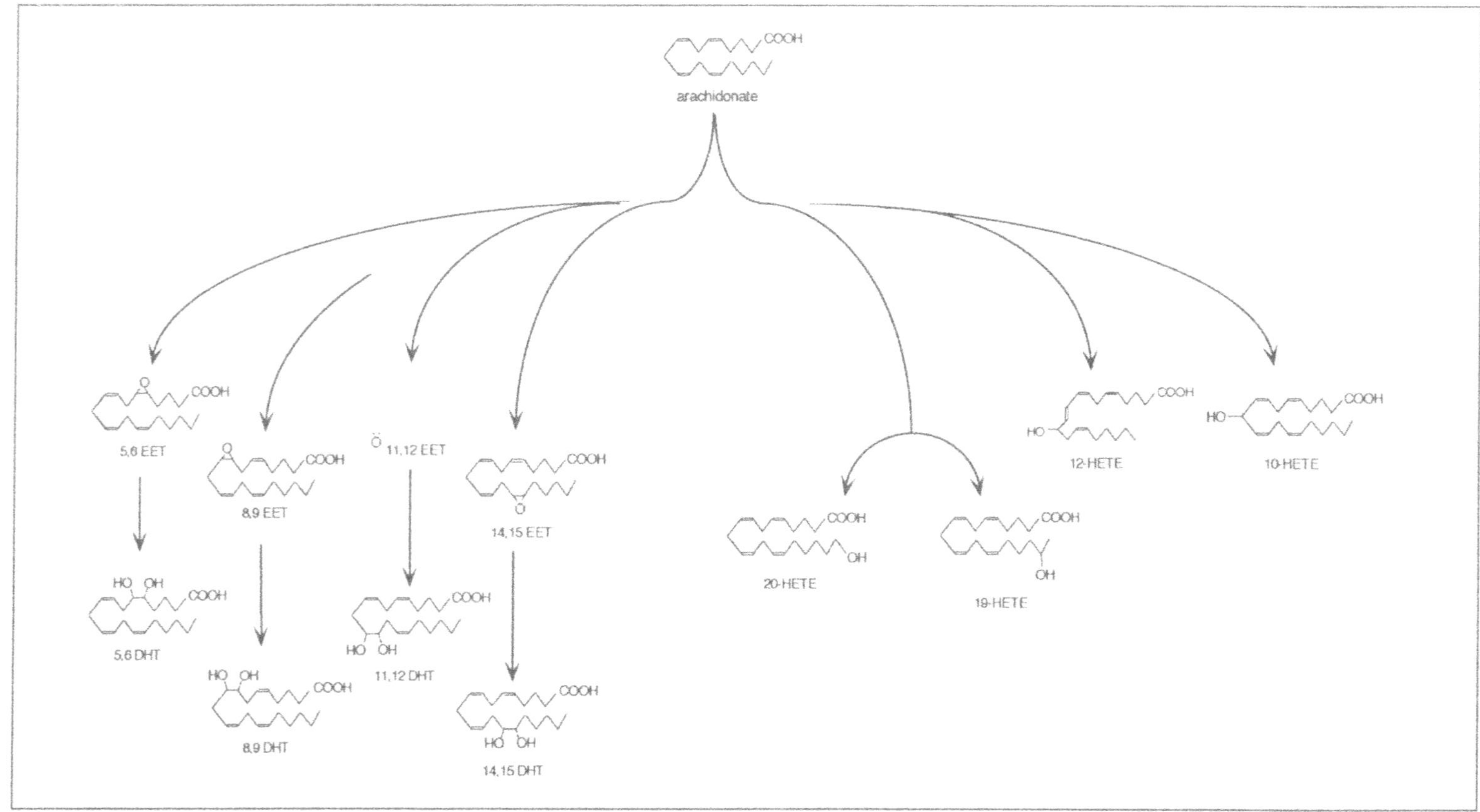

Fig. 4.26. Arachidonate metabolism via the cytochrome P_{450} pathway.

the corresponding dihydroxyeicosatrienoic acids (DHTs). Second, the conversion of arachidonate to a number of hydroxyeicosatetranoic acids (HETEs) occurs.

Cytochrome P_{450} isoforms that convert arachidonic acid into EETs are referred to as arachidonate epoxygenases and have been purified to homogeneity from human liver and rat kidney.[150,151] A complementary DNA encoding the rat kidney arachidonate epoxygenase has also been cloned and characterized at the molecular level. Sequence analysis showed that this complementary DNA encodes a protein of 494 amino acids that is virtually identical to rat liver cytochrome 2C23, and whose mRNA is highly expressed in rat kidney and liver, but not in heart and brain.[152] Like other cytochrome P_{450} isoforms, arachidonate epoxygenase activity can be induced by a variety of drugs, such as phenobarbital.[153] Several cytochrome P450 inhibitors are in experimental use. These include SKF-525A, clotrimazole and 7-ethoxyresorufin.

Three main classes of HETEs derive from the cytochrome P_{450}-catalyzed oxygenation of arachidonic acid (Fig. 4.26): (1) HETEs containing a conjugated diene group, such as 12-HETE or 5-HETE. These are similar to those produced via the lipoxygenase pathways, but are usually formed in a non-stereoselective manner;[149] (2) the ω and ω-1 HETEs, 20-HETE and 19-HETE;[149] (3) the bis-allylic HETEs, 7-HETE, 10-HETE and 13-HETE.[154]

Epoxygenase metabolites of arachidonic acid have potent pharmacological effects on kidney and vascular functions. Particularly well studied is the ability of some of these products to inhibit ion transport, reduce renin activity, and increase water and electrolyte excretion in kidney.[155,156] Epoxygenase metabolism of arachidonic acid has also been demonstrated in brain,[157,158] where it could participate in mediating certain intracellular responses of hypothalamic neurons to the modulatory neurotransmitter, dopamine.[158]

References

1. Smith WL, Lands WEM. Prostaglandins and arachidonate metabolites. In: Colowick SP, Kaplan NO, eds. Methods in Enzymology. Volume 86 New York: Academic Press, 1982.

2. Murphy RC, Fitzpatrick FA. Arachidonate related lipid mediators. In: Abelson JN, Simon MI, eds. Methods in Enzymology. Volume 187 San Diego: Academic Press, 1990.
3. Touchstone JC, Dobbins MF. Practice of Thin-Layer Chromatography. New York: John Wiley and Sons, 1983.
4. Henderson RJ, Tocher DG. Thin-layer chromatography. In: Hamilton RJ, Hamilton S, eds. Lipid Analysis. A Practical Approach. New York: Oxford University Press, 1992: 65-111.
5. Hamberg M, Svensson J, Samuelsson B. Thromboxanes: A new group of biologically active compounds derived from prostaglandin endoperoxides. Proc Natl Acad Sci USA 72:2994-2998.
6. Gréen K, Samuelsson B. Prostaglandin and related factors: XIX. Thin-layer chromatography of prostaglandins. J Lip Research 1964; 5:117-120.
7. Christie WWC. HPLC and Lipids. Oxford:Pergamon Press, 1987.
8. Sewell PA. High-performance liquid chromatography. In: Hamilton RJ, Hamilton S, eds. Lipid Analysis: A Practical Approach. New York: Oxford University Press, 1992: 153-203.
9. Murphy RC, Hammarström S, Samuelsson B. Leukotriene C: A slow-reacting substance from murine mastocytoma cells. Proc Natl Acad Sci USA 1979; 76:4275-4279.
10. Lewis RA, Austen KF, Drazen JM, Clark DA, Marfat A, Corey EJ. Slow-reacting substances of anaphylaxis: identification of leukotriene C-1 and D from human and murine sources. Proc Natl Acad Sci USA 1980; 77:3710-3714.
11. Serhan CN, Hamberg M, Samuelsson B. Lipoxins: novel series of biologically active compounds formed from arachidonic acid in human leukocytes. Proc Natl Acad USA 1984; 81:5335-5339.
12. Yu W, Powell WS. Analysis of leukotrienes, lipoxins, and monooxygenated metabolites of arachidonic acid by reversed-phase high-pressure liquid chromatography. Anal Biochem 1995; 226:241-251.
13. Kramer HJ, Stevens J, Seeger W. Analysis of 2- and 3-series prostanoids by post-HPLC ELISA. Anal Biochem 1993; 214: 535-543.
14. Hofer G, Bieglmayer C, Kopp B, Janisch H. Measurement of eicosanoids in menstrual fluid by the combined use of high pressure chromatography and radioimmunoassay. Prostaglandins 1993; 45:413-426.
15. Demin P, Reynaud D, Pace-Asciak CR. Extractive derivatization of the 12-lipoxygenase products, hepoxilins, and related compounds into fluorescent anthryl esters for their complete high-performance liquid chromatography profiling in biological systems. Anal Biochem 1995; 226:252-255.

16. Watson JT. Introduction to Mass Spectrometry. New York: Raven Press, 1985.
17. Burlingame AL, Baillie TA, Russel DH. Mass spectrometry. Anal Chem 1992; 64:467R-502R.
18. Rosenfeld JM, Moharir Y, Hill R. Direct solid-phase isolation and oximation of prostaglandin E_2 from plasma and quantitation by gas chromatography with mass spectrometric detection in the negative-ion chemical ionization mode. Anal Chem 1992; 63:1536-1541.
19. Miyamoto T, Ogino N, Yamamoto S, Hayaishi O. Purification of prostaglandin endoperoxide synthetase from bovine vesicular gland microsomes. J Biol Chem 1976; 251:2629-2636.
20. Van den Ouderaa FJ, Buytenhek M, Nugteren DH, Van Dorp DA. Purification and characterization of prostaglandin endoperoxide synthetase from sheep vesicular glands. Biochim Biophys Acta 1977; 487:315-331.
21. Van den Ouderaa FJ, Buytenhek M, Slikkerveer FJ, Van Dorp DA. On the haemoprotein character of prostaglandin endoperoxide synthetase. Biochim Biophys Acta 1979; 572:29-42.
22. DeWitt DL, Smith WL. Primary structure of prostaglandin G/H synthase from sheep vesicular gland determined from the complementary DNA sequence. Proc Natl Acad Sci USA 1988; 85:1412-1416.
23. Merlie JP, Fagan D, Mudd J, Needleman P. Isolation and characterization of the complementary DNA for sheep seminal vesicle prostaglandin endoperoxide synthase (cyclooxygenase). J Biol Chem 1988; 263:3550-3553.
24. Yokohama C, Takai T, Tanabe T. Primary structure of sheep prostaglandin endoperoxide synthase deduced from cDNA sequence. FEBS Letters 1988; 231:347-351.
25. DeWitt DL, El-Harith EA, Kraemer SA, Andrews MJ, Yao EF, Armstrong RL, Smith WL. The aspirin and heme-binding sites of ovine and murine prostaglandin endoperoxide synthases. J Biol Chem 1990; 265:5192-5298.
26. Yokohama C, Tanabe T. Cloning of human gene encoding prostaglandin endoperoxide synthase and primary structure of the enzyme. Biochem Biophys Res Commun 1989; 165:888-894.
27. Picot D, Loll PJ, Garavito RM. The X-ray crystal structure of the membrane protein prostaglandin H_2 synthase-1. Nature 1994; 367:243-249.
28. Kujubu DA, Fletcher BS, Varnum BC, Lim RW, Herschman HR. TIS10, a phorbol ester tumor promoter-inducible mRNA from Swiss 3T3 cells, encodes a novel prostaglandin synthase/cyclooxygenase homologue. J Biol Chem 1991; 266:12866-12872.

29. Fletcher BS, Kujubu DA, Perrin DM, Herschman HR. Structure of the mitogen-inducible TIS10 gene and demonstration that the TIS10-encoded protein is a functional prostaglandin G/H synthase. J Biol Chem 1992; 267:4338-4344.
30. Kujubu DA, Herschman HR. Dexamethasone inhibits mitogen induction of the TIS10 prostaglandin synthase/cyclooxygenase gene. J Biol Chem 1992; 267:7991-7994.
31. O'Banion MK, Winn VD, Young DA. cDNA cloning and functional activity of a glucocorticoid-regulated inflammatory cyclooxygenase. Proc Natl Acad Sci USA 1992; 89:4888-4892.
32. Masferrer JL, Seibert K, Zweifel B, Needleman P. Endogenous glucocorticoid regulate an inducible cyclooxygenase enzyme. Proc Natl Acad Sci USA 1992; 89:3917-3921.
33. Yamagata K, Andreasson KI, Kaufmann WE, Barnes CA, Worley PF. Expression of a mitogen-inducible cyclooxygenase in brain neurons: regulation by synaptic acitivty and glucocorticoids. Neuron 1993; 11:371-386.
34. Smith WL, Marnett LJ. Prostaglandin endoperoxide synthase: structure and catalysis. Biochim Biophys Acta 1991; 1083:1-17.
35. Pace-Asciak CR, Smith WL. Enzymes in the biosynthesis and catabolism of the eicosanoids: prostaglandins, thromboxanes, leukotrienes and hydroxy fatty acids. In: Boyer PD, Krebs EG, eds. The Enzymes vol. 16, Orlando: Academic Press, 1983; 543-603.
36. Ferreira SH, Moncada S, Vane JR. Indomethacin and aspirin abolish prostaglandin release from the spleen. Nature [New Biology] 1971; 231:237-239.
37. Smith JB, Willis AL. Aspirin selectively inhibits prostaglandin production in human platelets. Nature [New Biology] 1971; 231:235-237.
38. Goetzl EJ, An S, Smith WL. Specificity of expression and effects of eicosanoid mediators in normal physiology and human disease. FASEB J 1995; 9:1051-1058.
39. DeWitt DL. Prostaglandin endoperoxide synthase: regulation of enzyme expression. Biochim Biophys Acta 1991; 1083:121-134.
40. Masferrer JL, Zweifel BS, Manning PT, Hauser SD, Leahy KM, Smith WG, Isakson PC, Seibert K. Selective inhibition of inducible cyclooxygenase 2 in vivo is anti-inflammatory and nonulcerogenic. Proc Natl Acad Sci USA 1994; 91:3228-3232.
41. Seibert K, Zhang Y, Leahy KM, Hauser SD, Masferrer JL, Perkins W, Lee L, Isakson PC. Pharmacological and biochemical demonstration of the role of cyclooxygenase 2 in inflammation and pain. Proc Natl Acad Sci USA 1994; 91:12013-12017.

42. Morham SG, Langenbach R, Loftin CD et al. Prostaglandin synthase 2 gene disruption causes severe renal pathology in the mouse. Cell 1995; 83:473-482.
43. Langenbach R, Morham SG, Tiano HF et al. Prostaglandin synthase 1 gene disruption in mice reduces arachidonic acid-induced inflammation and indomethacin-induced gastric ulceration. Cell 1995; 83:483-492.
44. DeWitt D, Smith WL. Yes, but do they still get headaches? Cell 1995; 83:345-348.
45. Shimizu T, Yamamoto S, Hayaishi O. Purification and properties of prostaglandin D synthetase from rat brain. J Biol Chem 1979; 259:5222-5228.
46. Urade Y, Fujimoto N, Hayaishi O. Purification and characterization of rat brain prostaglandin D synthetase. J Biol Chem 1984; 260:12410-12415.
47. Urade Y, Nagata A, Suzuki Y, Fuji Y, Hayaishi O. Primary structure of rat brain prostaglandin D synthetase deduced from cDNA sequence. J Biol Chem 1989; 264:1041-1045.
48. Nagata A, Suzuki Y, Igarashi M, Eguchi N, Toh H, Urade Y, Hayaishi O. Human brain prostaglandin D synthase has been evolutionarily differentiated from lipophilic-ligand carrier proteins. Proc Natl Acad USA 1991; 88:4020-4024.
49. Shimizu T, Mizuno N, Amano T, Hayaishi O. Prostaglandin D_2, a neuromodulator. Proc Natl Acad USA 1979; 76:6231-6234.
50. Hayaishi O. Sleep-wake regulation of prostaglandins D_2 and E_2. J Biol Chem 1988; 263:14593-14596.
51. Christ-Hazelhof E, Nugteren DH. Isolation of PGH-PGD isomerase from rat spleen. In: Colowick SP, Kaplan NO, eds. Methods in Enzymology. New York: Academic Press, 1982; 86 :77-91.
52. Manning DC. Hyperalgesia, cutaneous chemical mechanisms. In: Adelman G, Neuroscience Year, Boston: Birkhauser, 1989:77-79.
53. Ujihara M, Tsuchida S, Satoh K, Sato K, Urade Y. Biochemical and immunological demonstration of prostaglandin D_2, E_2, and $F_{2\alpha}$ formation from prostaglandin H_2 by various rat glutathione S-transferase isozymes. Arch Biochem Biophys 1988; 264:428-437.
54. Kikawa Y, Narumiya S, Fukushima M, Wakatsuka H, Hayaishi O. 9-Deoxy-$\Delta 9,\Delta^{12}$-13,14-dihydro prostaglandin D_2, a metabolite of prostaglandin D_2 formed in human plasma. Proc Natl Acad USA 1984; 81:1317-1321.
55. Forman BM, Tontonoz P, Chen J, Brun RP, Spiegelman BM, Evans RM. 15-Deoxy-$\Delta^{12,14}$ prostaglandin J_2 is a ligand for the adipocyte determination factor PPARγ. Cell 1995; 83:803-812.

56. Kliewer SA, Lenhard JM, Willson TM, Patel I, Morris DC, Lehmann J. A prostaglandin J2 metabolite binds peroxisome proliferator-activated receptor γ and promotes adipocyte differentiation. Cell 1995; 83:813-819.
57. Wong PYK. Purification of PGD2 11-ketoreductase from rabbit liver. In: Colowick SP, Kaplan NO, eds. Methods in Enzymology. New York: Academic Press, 1982; 86 :117-125.
58. Wolfe LS, Pellerin L. Arachidonic acid metabolites in the rat and human brain. Annals NY Acad Sci 1988; 559:374-381.
59. Shimizu T, Watanabe K, Tokumoto H, Hayaishi O. Isolation of $NADP^+$-dependent PGD_2-specific 15-hydroxyprostaglandin dehydrogenase from swine brain. In: Colowick SP, Kaplan NO, eds. Methods in Enzymology. New York: Academic Press, 1982; 86:147-155.
60. Ogino M, Miyamoto T, Yamamoto S, Hayaishi O. Prostaglandin endoperoxide E isomerase from bovine vesicular gland microsomes, a glutathione-requiring enzyme. J Biol Chem 1977; 252:890
61. Moonen P, Buytenhek M, Nugteren DH. Purification of PGH-PGE isomerase from sheep vesicular gland. In: Colowick SP, Kaplan NO, eds. Methods in Enzymology. New York: Academic Press, 1982; 86:84-91.
62. Tanaka Y, Ward SL, Smith WL. Immunochemical and kinetic evidence for two different prostaglandin H-prostaglandin E isomerases in sheep vesicular gland microsomes. J Biol. Chem 1987; 262:1374-1381.
63. Anggard E, Samuelsson B. Prostaglandins and related factors. 28. Metabolism of PGE_1 in guinea pig lung: the structures of two metabolites. J Biol Chem 1964; 239:4097-4102.
64. Pace-Asciak CR, Carrara MC, Domazet Z. Identification of the major urinary metabolites of 6-keto-$PGF_{1\alpha}$ in the rat. Biochem Biophys Res Commun 1977; 78:115-121.
65. Sun FF, Taylor BM. Metabolism of prostacyclin in rat. Biochemistry 1978; 17:4096-4101.
66. Roberts LJ, Sweetman BJ, Payne NA, Oates JA. Metabolism of thromboxane B_2 in man. Identification of the major urinary metabolites. J Biol Chem 1977; 252:7415-7417.
67. Lee SC, Levine L. Prostaglandin metabolism. II. Identification of two 15-hydroxyprostaglandin dehydrogenase isoforms. J Biol Chem 1975; 250:548-552.
68. Murphy RC, Fitzgerald GA. Current approaches to estimation of eicosanoid formation in vivo. Advances in Prostaglandin, Thromboxane and Leukotriene Res 1994; 22:341-348.
69. Oliw EH, Fahlstadius P, Hamberg M. Isolation and biosynthesis of 20-hydroxyprostaglandins E1 and E2 in ram seminal fluid. J Biol Chem 1986; 261:9216-9221.

70. Moncada S, Gryglewski R, Bunting S, Vane JR. An enzyme isolated from arteries transforms prostaglandin endoperoxides to an unstable substance that inhibits platelet aggregation. Nature 1976; 263:663-665.
71. Pace-Asciak CR, Wolfe LS. A novel prostaglandin derivative formed from arachidonic acid by rat stomach homogenates. Biochemistry 1971; 10:3657-3664.
72. Wise H, Jones RL. Focus on prostacyclin and its novel mimetics. Trends Pharmacol Sci 1996; 17:17-21.
73. Shen RF, Tai HH. Immunoaffinity purification and characterization of thromboxane synthase from porcine lung. J Biol Chem 1986; 261:11592-11599.
74. Haraund M, Ullrich V. Isolation and characterization of thromboxane synthase from human platelets as a cytochrome P-450 enzyme. J Biol Chem 1986; 260:15059-15067.
75. Yokoyama C, Miyata A, Ihara A, Ullrich V, Tanabe T. Molecular cloning of human platelet thromboxane A synthase. Biochem Biophys Res Comm 1991; 178:1479-1484.
76. Zhang L, Chase MB, Shen RF. Molecular cloning and expression of murine thromboxane synthase. Biochem Biophys Res Comm 1993; 194:741-748.
77. Tanabe T, Yokoyama C, Miyata A, Ihara A, Kosaka T, Suzuki K, Nishikawa Y, Yoshimoto T, Yamamoto S, Nusing R, et al. Molecular cloning and expression of human thromboxane synthase. J Lipid Mediators 1993; 6:139-144.
78. Shen RF, Baek SJ, Lee KD, Zhang L, Fleischer T. Genomic organization and regulation of the human thromboxane synthase gene. Advances in Prostaglandin, Thromboxane and Leukotriene Res 1995; 23:137-139.
79. Wang LH, Tazawa R, Lang AQ, Wu KK. Alternate splicing of human thromboxane synthase mRNA. Archives Biochem Biophys 1994; 315:273-278.
80. Ruan KH, Li P, Kulmacz RJ, Wu KK. Characterization of the structure and membrane interaction of NH_2-terminal domain of thromboxane A_2 synthase. J Biol Chem 1994; 269:20938-20942.
81. Vermylen J, Deckmyn H. Thromboxane synthase inhibitors and receptor antagonists. Cardiovascular Drugs and Therapy 1992; 6:29-33.
82. Kurosawa M. Role of thromboxane A_2 synthase inhibitors in the treatment of patients with bronchial asthma. Clinical Therapeutics 1995; 17:2-11.
83. Ackerley N, Brewster AG, Brown GR, Clarke DS, Foubister AJ, Griffin SJ, Hudson JA, Smithers MJ, Whittamore PR. A novel approach to dual-acting thromboxane receptor antagonist/synthase inhibitors based on the link of 1,3 dioxane-thromboxane receptor antagonist and thromboxane synthase inhibitors. J Med Chem 1995; 38:1608-1628.

84. Watanabe K, Yoshida R, Shimizu T, Hayaishi O. Enzymatic formation of prostaglandin $F_{2\alpha}$ from prostaglandin H_2 and D_2. Purification and properties of prostaglandin F synthetase from bovine lung. J Biol Chem 1985; 260:7035-7041.
85. Watanabe K, Fuji Y, Nakayama K, Ohkubo H, Kuramitsu S, Kagamiyama H, Nakanishi S, Hayaishi O. Structural similarity of the bovine lung prostaglandin F synthetase to lens ε-crystallin of the European common frog. Proc Natl Acad Sci USA 1988; 85:11-15.
86. Hayaishi H, Fuji Y, Watanabe K, Urade Y, Hayaishi O. Enzymatic conversion of prostaglandin H_2 to prostaglandin $F_{2\alpha}$ by aldehyde reductase from human liver: comparison to the prostaglandin F synthetase from bovine lung. J Biol Chem 1989; 264:1036-1040.
87. Morrow JD, Hill KE, Burk RF, Nammour TM, Badr KF, Roberts LJ. A series of prostaglandin F_2-like compounds are produced in vivo in humans by a non-cyclooxygenase, free radical-catalyzed mechanism. Proc Natl Acad Sci USA 1990; 87:9383-9387.
88. Morrow JD, Awad JA, Boss HJ, Blair IA, Roberts LJ. Non-cyclooxygenase-derived prostanoids (F_2-isoprostanes) are formed in situ on phospholipids. Proc Natl Acad Sci USA 1992; 89:10721-10725.
89. Kayganich-Harrison KA, Rose DM, Murphy RC, Morrow JD, Roberts LJ. Collision-induced dissociation of F_2-isoprostanes-containing phospholipids. J Lipid Res 1993; 34:1229-1235.
90. Takahashi K, Nammour TM, Fukunaga M, Ebert J, Morrow JD, Roberts LJ , Hoover RL, Badr KF. Glomerular actions of a free radical-generated novel prostaglandin, 8-epi-prostaglandin $F_{2\alpha}$, in the rat. J Clin Invest 1992; 90:136-141.
91. Pratico D, Lawson JA, Fitzgerald GA. Cyclooxygenase-dependent formation of the isoprostane, 8-epi prostaglandin $F_{2\alpha}$. J Biol Chem 1995; 270:9800-9808.
92. Yamamoto S. Mammalian lipoxygenases: molecular structures and functions. Biochim Biophys Acta 1992; 1128:117-131.
93. Ford-Hutchinson AW, Gresser M, Young RN. 5-Lipoxygenase. Annu Rev Biochem 1994; 63:383-417.
94. Balcarek JM, Theisen TW, Cook MN, Varricchio A, Hwang SM, Strohsacker MW, Crooke ST. Isolation and characterization of a cDNA encoding rat 5-lipoxygenase. J Biol Chem 1988; 263:13937-13941.
95. Rouzer CA, Rands E, Kargman S, Jones RE, Register RB, Dixon RAF. Characterization of cloned human leukocyte 5-lipoxygenase expressed in mammalian cells. J Biol Chem 1988; 263:10135-10140.

96. Dixon RAF, Jones RE, Diehl RE, Bennet CD, Kargman S, Rouzer CA. Cloning of the cDNA for human 5-lipoxygenase. Proc Natl Acad Sci USA 1988; 85:416-420.
97. Funk CD, Gunne H, Steiner H, Izumi T, Samuelsson B. Native and mutant 5-lipoxygenase expression in a baculovirus insect cell system. Proc Natl Acad Sci USA 1989; 86:2592-2596.
98. Funk CD, Hoshiko S, Matsumoto T, Rådmark O, Samuelsson B. Characterization of the human 5-lipoxygenase gene. Proc Natl Acad Sci USA 1989; 86:2587-2591.
99. Nakamura M, Matsumoto T, Noguchi M, Yamashita I, Noma M. Expression of cDNA encoding human 5-lipoxygenase under control of the *STA1* promoter in *Saccharomyces cerevisiae*. Gene 1990; 89:231-237.
100. Percival MD. Human 5-lipoxygenase contains an essential iron. J Biol Chem 1991; 266:10058-10061.
101. Lammers CH, Schweitzer P, Facchinetti P, Arrang JM, Madamba SG, Siggins GR, Piomelli D. Arachidonate 5-lipoxygenase and its activating protein: prominent hippocampal expression and role in somatostatin signaling. J Neurochem 1996; 66:147-152.
102. Nicholson DW, Ali A, Vaillancourt JP, Calaycay JR, Mumford RA, Zamboni RJ, Ford-Hutchinson A. Purification to homogeneity and the N-terminal sequence of human leukotriene C_4 synthase: a homodimeric glutathione S-transferase composed of 18-kDa subunits. Proc Natl Acad Sci USA 1993; 90:2015-2119.
103. Welsch DJ, Creely DP, Hauser SD, Methis KJ, Krivi GG, Isakson PC. Molecular cloning and expression of human leukotriene C_4 synthase. Proc Natl Acad Sci USA 1994; 91:9745-9749.
104. An S, Schmid FJ, Campbell BJ. Molecular cloning of sheep lung dipeptidase: a glycosyl phosphatidylinositol-anchored ectoenzyme that converts leukotriene D_4 to leukotriene E_4. Biochim Biophys Acta 1994; 1226:337-340.
105. Rådmark O, Haeggström J. Properties of leukotriene A_4 hydrolase. Advances Prostaglandin Thromboxane Res 1990; 20:35-45.
106. Ohishi N, Minami M, Kobayashi J, Seyama Y, Hata J, Yotsumoto H, Takaku F, Shimizu T. Immunological quantitation and immunohistochemical localization of leukotriene A_4 hydrolase in guinea pig tissues. J Biol Chem 1990; 265:7520-7525.
107. Lindgren JA, Edenius C. Transcellular biosynthesis of the leukotrienes and lipoxins via leukotriene A_4 transfer. Trends Pharmacol Sci 1993; 14:351-353.
108. Orning L, Krivi G, Fitzpatrick FA. Leukotriene A_4 hydrolase. Inhibition by bestatin and intrinsic aminopeptidase activity establish its functional resemblance to metallohydrolase enzymes. J Biol Chem 1991; 266:1375-1378.

109. Minami M, Ohishi N, Mutoh H, Tsumi T, Bito H, Wada Y, Seyama Y, Toh H, Shimizu T. Leukotriene A_4 hydrolase is a zinc-containing aminopeptidase. Biochem Biophys Res Commun 1990; 173:620-626.
110. Medina JF, Wetterholm A, Rådmark O, Shapiro R, Haeggström JZ, Vallee BL, Samuelsson B. Leukotriene A_4 hydrolase: determination of the three zinc-binding ligands by site-directed mutagenesis and zinc analysis. Proc Natl Acad Sci USA 1991; 88:7620-7624.
111. Hay DWP, Torphy TJ, Undem BJ. Cysteinyl leukotrienes in asthma: old mediators up to new tricks. Trends Pharmacol Sci 1995; 16:304-309.
112. Chen XS, Sheller JR, Johnson EN, Funk CD. Role of leukotrienes revealed by targeted disruption of the 5-lipoxygenase gene. Nature 1994; 372:179-182.
113. McMillan RM, Walker ERH. Designing therapeutically effective 5-lipoxygenase inhibitors. Trends Pharmacol Sci 1992; 13:323-330.
114. Gillard J, Ford-Hutchinson AW, Chan C et al. L-663,536 (MK-866) (3-[1-(4-chlorobenzyl)-3-t-butyl-thio-5-isopropylindol-2-yl]-2,2-dimethylpropanoic acid), a novel, orally active leukotriene biosynthesis inhibitor. Canadian J Physiol Pharmacol 1989; 67:456-464.
115. Rouzer CA, Ford-Hutchinson AW, Morton HE, Gillard JW. MK-886, a potent and specific leukotriene biosynthesis inhibitor blocks and reverses the membrane association of 5-lipoxygenase in ionophore-challenged leukocytes. J Biol Chem 1990; 265: 1436-1442.
116. Wetterlholm A, Haeggström JZ, Samuelsson B, Yuan B, Munoz B, Wong CH. Potent and selective inhibitors of leukotriene A_4 hydrolase: effects on purified enzyme and human polymorphonuclear leukocytes. J Pharmacol Exper Ther 1995; 275:31-37.
117. Funk CD, Furci L, Fitzgerald GA. Molecular cloning, primary structure, and expression of the human platelet/erythroleukemia cell 12-lipoxygenase. Proc Natl Acad Sci USA 1990; 87:5638-5642.
118. Izumi T, Hoshiko S, Rådmark O, Samuelsson B. Cloning of the cDNA for human 12-lipoxygenase. Proc Natl Acad Sci USA 1990; 87:7477-7481.
119. Yoshimoto T, Sukuki H, Yamamoto S, Takai T, Yokoyama C, Tanabe T. Cloning and sequence analysis of the cDNA for arachidonate 12-lipoxygenase of porcine leukocytes. Proc Natl Acad Sci USA 1990; 87:2142-2146.
120. Funk CD, Funk LB, Fitzgerald GA, Samuelsson B. Characterization of human 12-lipoxygenase gene. Proc Natl Acad Sci USA 1992; 89:3962-3966.

121. Arakawa T, Oshima T, Kishimoto K, Yoshimoto T, Yamamoto S. Molecular structure and function of the porcine 12-lipoxygenase gene. J Biol Chem 1992; 267:12188-12191.
122. Watanabe T, Medina JF, Haeggström JZ, Rådmark O, Samuelsson B. Molecular cloning of a 12-lipoxygenase from rat brain. Eur J Biochem 1993; 212:605-612.
123. Funk CD. Molecular biology of the eicosanoid field. Progress Nucleic Acid Res Mol Biol 1993; 45:67-98.
124. Hada T, Hagiya H, Suzuki H et al. Arachidonate 12-lipoxygenase of rat pineal glands: catalytic properties and primary structure deduced from its cDNA. Biochim Biophys Acta 1994; 1211:221-228.
125. Natarajan R, Gonzales N, Hornsby PJ, Nadler J. Mechanism of angiotensin II-induced proliferation in bovine adrenocortical cells. Endocrinology 1992; 131:1174-1180.
126. Timar J, Chen YQ, Liu B, Baza R, Taylor JD, Honn K. The lipoxygenase metabolite 12(S)-HETE promotes $\alpha_{IIb}\beta_3$ integrin-mediated tumor-cell spreading on fibronectin. Int J Cancer 1992; 52:594-603.
127. Pace-Asciak CR, Granström E, Samuelsson B. Arachidonic acid epoxides. Isolation and structure of two hydroxy epoxide intermediates in the formation of 8,11,12- and 10,11,12-trihydroxyeicosatrienoic acids. J Biol Chem 1983; 258:6835-6840.
128. Piomelli D, Shapiro E, Zipkin R, Schwartz JH, Feinmark SJ. Formation and action of 8-hydroxy-11,12-epoxy-5,9,14-icosatrienoic acid in *Aplysia*: a possible second messenger in neurons. Proc Natl Sci USA 1989; 86:1721-1725.
129. Pace-Asciak CR. Formation and metabolism of hepoxilin A_3 in the rat brain. Biochem Biophys Res Commun 1988; 151:493-498.
130. Carlen PL, Gurevich N, Wu PH, Su W-G, Corey EJ, Pace-Asciak CR. Actions of arachidonic acid and hepoxilin A_3 on mammalian hippocampal CA1 neurons. Brain Res 1989; 497:171-176.
131. Carlen PL, Gurevich N, Zhang L, Wu PH, Reynaud D, Pace-Asciak CR. Formation and electrophysiological actions of the arachidonic acid metabolites, hepoxilins, at nanomolar concentrations in rat hippocampal slices. Neuroscience 1994; 58:493-502.
132. Reynaud D, Delton I, Gharib A, Sarda N, Lagarde M, Pace-Asciak CR. Formation, metabolism and action of hepoxilin A_3 in the rat pineal gland. J Neurochem 1994; 62:126-133.
133. Margalit A, Livne AA. Lipoxygenase product controls the regulatory volume decrease of human platelets. Platelets 1991; 2:207-214.
134. Margalit A, Sofer Y, Grossman S, Reynaud D, Pace-Asciak CR, Livne AA. Hepoxilin A_3 is the endogenous lipid mediator opposing hypotonic swelling of intact human platelets. Proc Natl Sci USA 1993; 90:2589-2592.

135. Piomelli D, Feinmark SJ, Shapiro E, Schwartz JH. Formation and biological activity of 12-ketoeicosatetraenoic acid in the nervous system of *Aplysia*. J Biol Chem 1988; 263:16591-16596.
136. Murray JJ, Brash AR. Rabbit reticulocyte lipoxygenase catalyzes specific 12(S) and 15(S) oxygenation of arachidonate-containing phospholipids. Arch Biochem Biophys 1988; 265:514-531.
137. Kühn H, Belkner J, Wiesner R, Brash A. Oxygenation of biological membranes by the pure reticulocyte lipoxygenase. J Biol Chem 1990; 265:18351-18361.
138. Narumiya S, Salmon SA, Cottee FH, Weatherly BC, Flower RJ. Arachidonic acid 15-lipoxygenase from rabbit peritoneal polymorphonuclear leukocytes. Partial purification and properties. J Biol Chem 1981; 256:9583-9592.
139. Sigal E, Grunberger D, Craik CS, Caughey GH, Nadel JA. Arachidonate 15-lipoxygenase (ω-6 lipoxygenase) from human leukocytes. Purification and structural homology to other mammalian lipoxygenases. J Biol Chem 1988; 263:5328-5332.
140. Sigal E, Craik CS, Highland E, Grunberger D, Costello LL, Dixon RAF, Nadel JA. Molecular cloning and primary structure of human 15-lipoxygenase. Biochem Biophys Res Commun 1988; 157:457-464.
141. Fleming B, Thiele BJ, Chester J, O'Prey J, Jaetzki S, Aitken A, Anton IA, Rapoport SM. The complete sequence of the rabbit eryhtroid cell-specific 15-lipoxygenase mRNA: comparison of the predicted amino acid sequence of the erythrocyte lipoxygenase with other lipoxygenases. Gene 1989; 79:181-188.
142. Izumi T, Rådmark O, Jörnvall H, Samuelsson B. Purification of two forms of arachidonate 15-lipoxygenase from human leukocytes. Eur J Biochem 1991; 202:1231-1238.
143. Boyinton JC, Gaffney BJ, Amzel LM. The three-dimensional structure of an arachidonic acid 15-lipoxygenase. Science 1993; 260:1482-1486.
144. Levine JD, Lam D, Taiwo YO, Donatoni P, Goetzl EJ. Hyperalgesic properties of 15-lipoxygenase products of arachidonic acid. Proc Natl Sci USA 1986; 83:5331-5334.
145. Schewe T, Kühn H. Do 15-lipoxygenases have a common biological role? Trends Biochem Sci 1991; 16:369-373.
146. Glasgow WC, Eling TE. Epidermal growth factor stimulates linoleic acid metabolism in BALB/c 3T3 fibroblasts. Mol Pharmacol 1990; 38:503-510.
147. Glasgow WC, Afshari CA, Barrett JC, Eling TE. Modulation of the epidermal growth factor mitogenic response by metabolites of linoleic and arachidonic acid in Syrian hamster embryo fibroblasts. Differential effects in tumor suppressor gene (+) and (-) phenotypes. J Biol Chem 1992; 267:10771-10779.

148. Serhan CN. Lipoxin biosynthesis and its impact in inflammatory and vascular events. Biochim Biophys Acta 1994; 1212:1-25.
149. McGiff JC. Cytochrome P_{450} metabolism of arachidonic acid. Annu Rev Pharmacol Toxicol 1991; 31:339-369.
150. Schwartzman ML, Davis KL, McGiff JC, Levere RD, Abraham NG. Purification and characterization of cytochrome P_{450}-dependent arachidonic acid epoxygenase from human liver. J Biol Chem 1988; 263:2536-2542.
151. Laethem RM, Laethem CL, Koop DR. Purification and properties of a cytochrome P450 arachidonic acid epoxygenase from rabbit renal cortex. J Biol Chem 1992; 267:5552-5559.
152. Karara A, Makita K, Jacobson HR, Falck JR, Guengerich FP, Dubois RN, Capdevila JH. Molecular cloning, expression and enzymatic characterization of the rat kidney cytochrome P_{450} arachidonic acid epoxygenase. J Biol Chem 1993; 268:13565-13570.
153. Capdevila JH, Karara A, Waxman DJ, Martin MV, Falck JR, Guengerich FP. Cytochrome P_{450} enzyme-specific control of the regio- and enantiofacial selectivity of the microsomal arachidonic acid epoxygenase. J Biol Chem 1990; 265:10865-10871.
154. Brash AR, Boeglin WE, Capdevila JH, Yeola S, Blair IA. 7-HETE, 10-HETE, and 13-HETE are major products of NADPH-dependent arachidonic acid metabolism in rat liver microsomes: analysis of their stereochemistry, and the stereochemistry of their acid-catalyzed rearrangement. Arch Biochem Biophysics 1995; 321:485-492.
155. Quilley CP, McGiff JC. Isomers of 12-hydroxy-5,8,10,14-eicosatetraenoic acid reduce renin activity and increase water and electrolyte excretion. J Pharmacol Exp Therap. 1990; 254:774-780.
156. Escalante B, Erlji D, Falck JR, McGiff JC. Effect of cytochrome P450 arachidonate metabolites on ion transport in rabbit kidney loop of Henle. Science 1991; 251:799-802.
157. Amrutesh SC, Falck JR, Ellis EF. Brain synthesis and cerebrovascular actions of epoxygenase metabolites of arachidonic acid. J Neurochem 1992; 58:503-510.
158. Junier MP, Dray F, Blair I, Capdevila J, Dishman E, Falck JR, Ojeda SR. Epoxygenase products of arachidonic acid are endogenous constituents of the hypothalamus involved in D_2 receptor-mediated, dopamine-induced release of somatostatin. Endocrinology 1990; 126:1534-1540.

CHAPTER 5

THE EICOSANOIDS IN CELLULAR SIGNALING

When a hormone or a neurotransmitter binds to a receptor on the membrane of a target cells, it triggers the formation of second messengers that are responsible for translating receptor occupancy into cellular responses. For example, the binding of dopamine to D_1-type receptors on neurons stimulates the activity of the enzyme adenylate cyclase, which catalyzes the conversion of ATP into cyclic AMP. This second messenger, in turn, binds to and activates a specific protein kinase, protein kinase A, which phosphorylates select intracellular proteins. Phosphorylation modifies the biological activity of these proteins and forms the basis for many physiological effects of dopamine in the central nervous system.[1]

This model of transmembrane signaling assumes that the range of action of a second messenger is confined to the intracellular environment. In support of this view, most signaling molecules—cyclic AMP, cyclic GMP, calcium, inositoltriphosphate, and diacylglycerol—produce their effects by binding to protein receptors located within the cell, whether they be protein kinases, protein phosphatases, calcium-binding proteins or ion channels. Such a model does not account, however, for all known forms of signal transduction.

As we have noted in chapter 1, the eicosanoids provide a good example of an alternative, more complex mode of action. These compounds are synthesized within cells, and they produce many of their biological effects by interacting with intracellular target proteins. Like non-esterified arachidonate, the eicosanoids appear thus to fulfill the criteria defining a second messenger system—receptor-dependent formation and intracellular site of action. Yet, they are also able to cross cell membranes, diffuse through the extracellular space and bind to receptors present on the membrane of neighboring cells. Therefore, they are also considered local mediators, i.e., autacoids (Fig. 5.1).

The purpose of the present chapter is to provide an overview of these distinct and complementary roles of the arachidonate cascade, intracellular and transcellular. We will first consider a series of messenger functions that arachidonate metabolites may serve within cells. We will turn next to the transport mechanisms used by the eicosanoids to exit cells and to gain access to transmembrane receptors present on the surface of cells nearby. Finally, we will briefly discuss the molecular structures of the eicosanoid receptors, their transduction mechanisms and some of their physiological and pharmacological properties.

EICOSANOIDS AS SECOND MESSENGERS

Two examples will be provided here to illustrate the various intracellular signaling functions served by arachidonate metabolites. The first, is the promotion of adipocyte differentiation by the prostaglandin, 15-deoxy-$\Delta^{12,14}$-PGJ_2, which involves binding of this cyclooxygenase metabolite to a nuclear receptor, and stimulation of gene expression. The second example is the regulation of neuronal excitability by products of the 12-lipoxygenase pathway, which may involve two separate effects: the activation of potassium-selective membrane channels and the inhibition of calcium/calmodulin-dependent protein kinase activity.

Several additional intracellular actions have been described for the eicosanoids, including the regulation of the multifunctional protein kinase, PKC,[2,3] of ion pumps such as the Na^+/K^+ ATPase,[4] and of a variety of membrane ion channels in excitable cells from

---

Fig. 5.1. (Opposite page) Possible mechanisms of actions of the eicosanoids in cell signaling. The eicosanoids are synthesized within cells, where they can exert their effects by modulating the activity of ion channels, protein kinases and other target proteins. They can also be extruded into the extracellular medium by selective membrane transporters. Once outside, they can bind to and activate G protein-linked membrane receptors on neighboring cells.

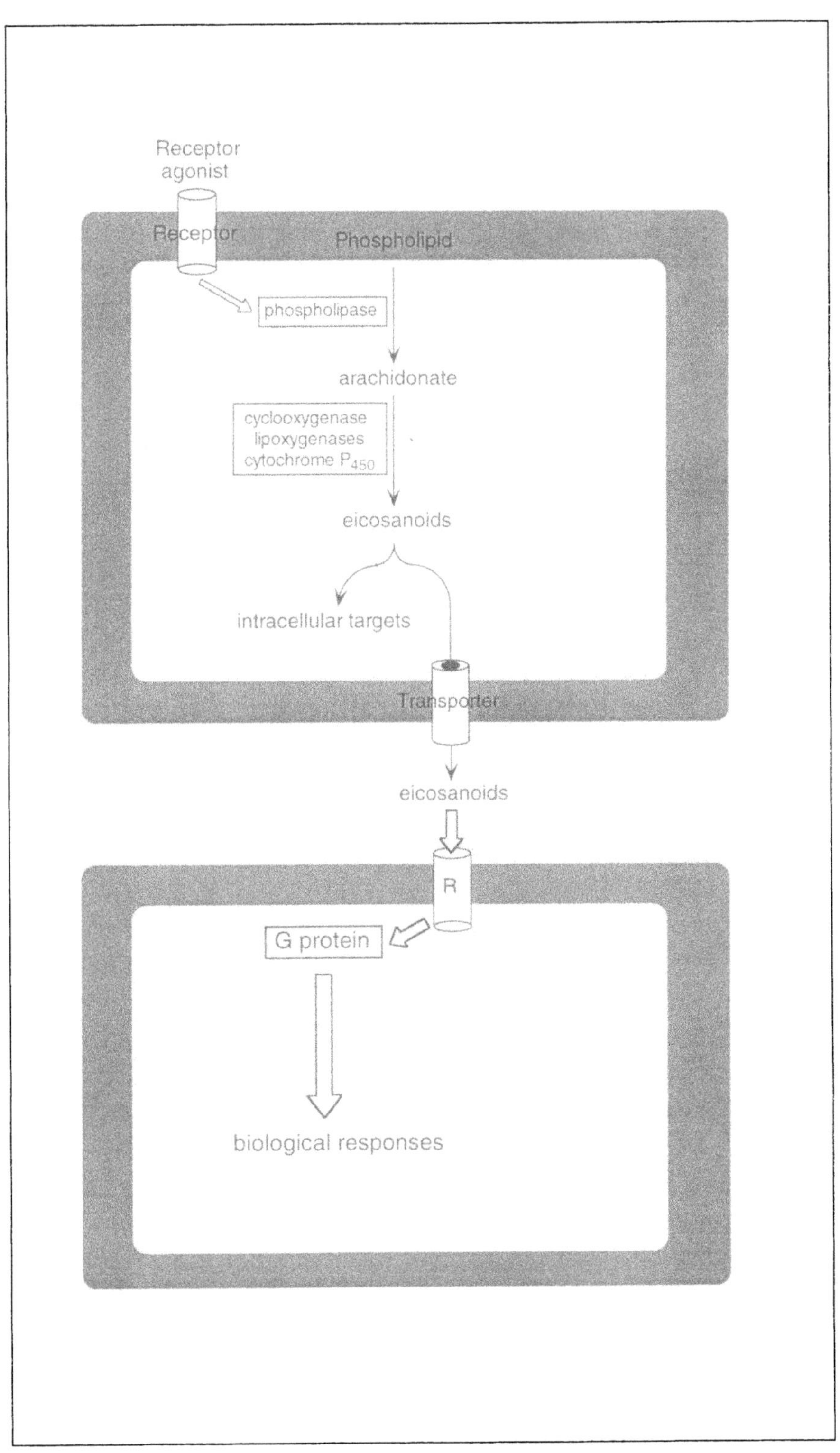
Receptor
agonist
Receptor
Phospholipid
phospholipase
arachidonate
cyclooxygenase
lipoxygenases
cytochrome P_{450}
eicosanoids
intracellular targets
Transporter
eicosanoids
R
G protein
biological responses

both vertebrate and invertebrate tissues.[5-7] We will not focus our attention on these findings here, but the interested reader may refer to two recent reviews that cover these subject matters.[8,9]

Induction of Cell Differentiation by PGD_2 Metabolites

The superfamily of nuclear receptor proteins is constituted by a group of transcription factors that are activated, in most cases, by hormonal ligands. For example, steroid hormones, thyroid hormones and retinoic acid bind selectively to receptors that belong to this superfamily. The hormone-receptor complexes interact with *cis*-acting elements present on the promoter sequences of target genes, initiating a series of transcription events that are responsible for the hormone's effects.[10]

The endogenous ligands for some of these nuclear receptors remain unknown, although pharmacological activators have been discovered. For example, drugs that stimulate peroxisome proliferation (e.g., the cholesterol-lowering agent, clofibrate) are selectively recognized by a unique class of nuclear receptors. Because their physiological agonist(s) are not known, these receptors have been called peroxisome proliferator-activated receptors (PPARs). Three mammalian PPAR subtypes have been identified thus far: PPARα, NUC1 (also called PPARδ) and PPARγ. The PPARγ subtype is abundantly expressed in adipocytes, and its pattern of expression correlates temporally with cell differentiation. In addition, ectopic expression of the PPARγ gene in fibroblasts results in the development of an adipocyte phenotype, suggesting that PPARγ may be a key player in adipocyte differentiation.[11,12]

How does PPARγ produce these responses? Like other members of the nuclear receptor superfamily, PPARγ contains a DNA-binding domain that recognizes hormone-responsive elements in the promoter regions of its target genes. Before binding to these sequences, however, PPARγ must first combine with the 9-*cis*-retinoic acid receptor, forming an active heterodimer (Fig. 5.2).[13]

The possibility that PPARγ participates in normal and pathological adipocyte differentiation has aroused considerable interest and has prompted an active search for endogenous molecules that may serve as ligands for this nuclear receptor. Along with drugs that stimulate peroxisome proliferation, PPARγ can also bind arachidonic acid, though with low affinity.[14-16] This observation led to the hypothesis that the endogenous PPARγ ligand(s) might be an arachidonate metabolite(s). Two groups of researchers independently set out to test this hypothesis and begun to screen a large number of commercially available eicosanoid compounds. Both groups arrived at the conclusion that biological activation of PPARγ can be very efficaciously achieved with several derivatives of the prostanoid, PGD_2, and particularly with PGJs[17,18] (for the chemistry of these compounds, see chapter 4).

In one of these studies, at least 10 different prostanoids were compared for their ability to activate PPARγ in CV1 cells, a cell line that expresses this receptor constitutively. PGD_2 and its two metabolites, PGJ_2 and Δ^{12}-PGJ_2, activated PPARγ by approximately 4-fold. Most effective in this respect was, however, a third PGD_2 metabolite, 15-deoxy-$\Delta^{12,14}$-PGJ_2, which caused a 14-fold enhancement in

Receptor subtype	Selective agonist	Selective antagonist	Tissue Distribution	Biological effects	Transduction system
BLT	LTB_4	LY255283	neutrophils, smooth muscle,	chemotaxis, chemokinesis, smooth muscle contraction	↑[Calcium]$_i$
cysLT_1	None	ICI1988615	smooth muscle, pulmonary epithelia, eosinophils	smooth muscle contraction, mucus secretion, eosinophil migration	↑[Calcium]$_i$
cysLT_2	None	None	smooth muscle	smooth muscle contraction	↑[Calcium]$_i$?

Fig. 5.2. Proposed mechanism of transcription activation by the peroxisome proliferator-activated receptor-γ(PPARγ). Upon binding to its ligand, the PPARγ forms a heterodimeric complex with a retinoic acid receptor. This complex interacts then with peroxisome proliferator-responsive elements located in the promoter sequences of target genes.

receptor activation. In contrast with PPARγ, the homologous PPAR subtype, NUC1, was not affected by any of the prostanoids tested.[17]

Before these experiments were reported, the general consensus in the eicosanoid field was that the prostanoids acted uniquely as local mediators. This was thought to be the case also for PGD_2, which is known to interact with its own cell surface receptor. As a result, when the cytosolic metabolism of PGD_2 to 15-deoxy-$\Delta^{12,14}$-PGJ_2 was discovered, it attracted only limited attention from cell biologists, although evidence that it could be of physiological importance in cell growth had been obtained.[19]

The results reported in the two PPARγ studies are therefore of great significance because they unequivocally demonstrate that the metabolism of PGD_2 can give rise to biologically active molecules that target nuclear transcription factors—an idea schematically depicted in Figure 5.3. These results open an unexpected, exciting avenue of prostanoid research. Moreover, they will likely prompt new experimental approaches for the identification of endogenous activators of other 'orphan' nuclear receptors, as well as for the development of drugs interfering with their functions.

Regulation of Neuronal Potassium Channels by 12-Lipoxygenase Products

Potassium-selective ion channels play a primary role in controlling the excitability of neuronal cells and the ability of these cells to release neurotransmitters. It is not surprising, therefore, that potassium channels are subject to a series of multiple regulatory mechanisms, which involve transmitter receptors, G proteins and soluble second messengers (see, for review, ref. 20).

Evidence that arachidonate metabolites participate in modulating potassium channel activity was first obtained in experiments carried out in the laboratory of James H. Schwartz, at the Center for Neurobiology and Behavior of Columbia University, by using the nervous system of the marine mollusk, *Aplysia californica*. *Aplysia* neural ganglia contain a relatively small number of large, easily identifiable neurons, which have been well characterized for

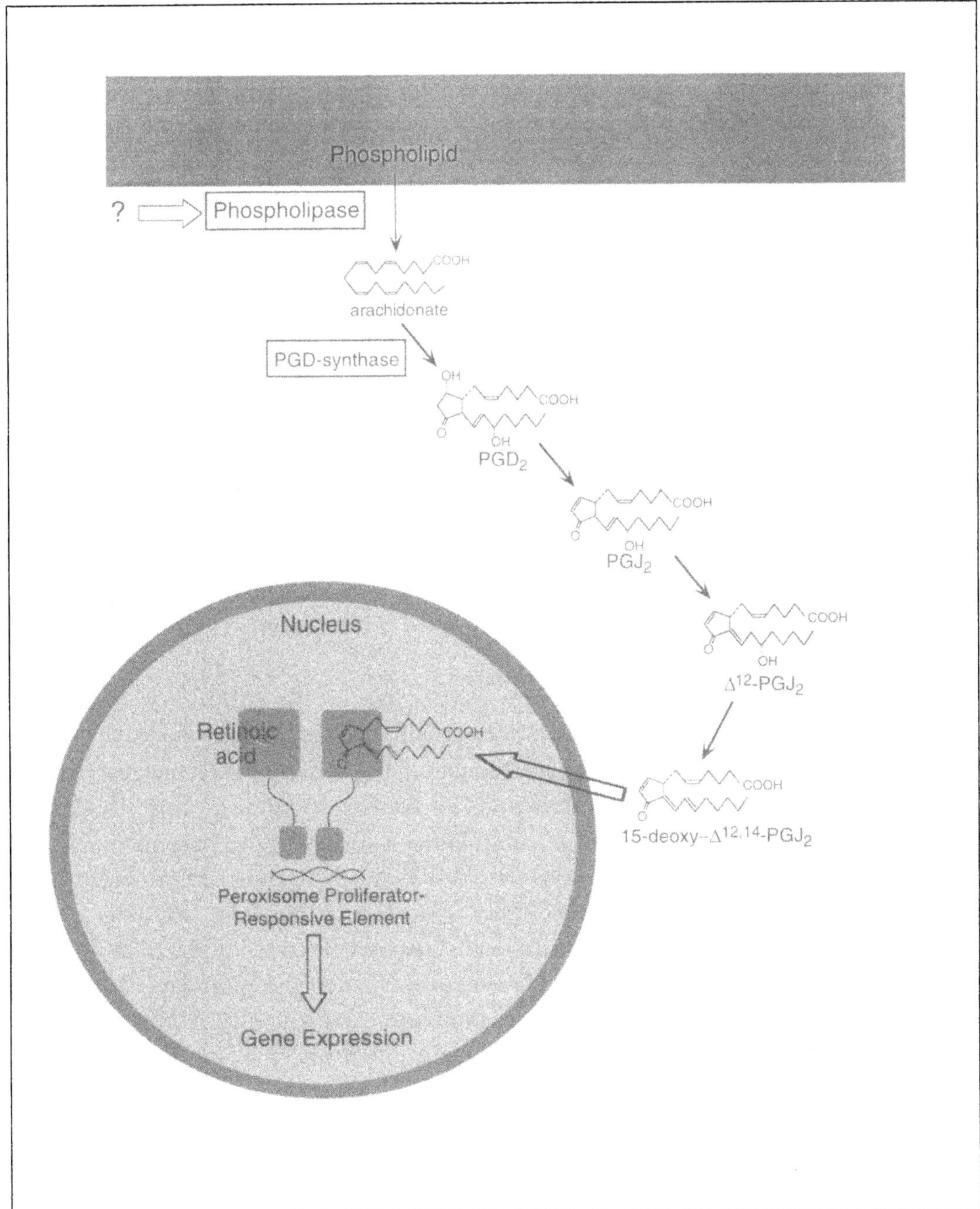

Fig. 5.3. Schematic model of the mechanism of formation and action of J-series prostaglandins (PGJ) in the activation of nuclear peroxisome proliferator-activated receptor-γ (PPARγ).

their neurotransmitter content, physiological effects and behavioral functions. Moreover, these neurons can be easily dissected out individually and kept in tissue culture for several days. These features make *Aplysia* neurons amenable to a variety of experimental manipulations that would be impossible in most vertebrate neurons.

When *Aplysia* neural ganglia are incubated in the presence of the modulatory neurotransmitter, histamine, they produce 12-lipoxygenase and 5-lipoxygenase metabolites of arachidonic acid. A similar effect is obtained when identified histaminergic neurons in the ganglia are stimulated electrically, which causes them to fire action potentials and to release histamine. In either case, the formation of arachidonate metabolites is prevented by treating the neural ganglia with cimetidine, a histamine receptor antagonist that selectively reduces two prominent actions of this biogenic amine— membrane hyperpolarization and presynaptic inhibition. Thus, physiological stimuli linked to presynaptic inhibition can elicit arachidonic acid metabolism, likely through the synaptic release of histamine and the activation of a receptor(s) coupled to phospholipase activity.[21]

Do arachidonate metabolites participate in the inhibitory actions of histamine? This question is difficult to address because the precise ionic conductances regulated by this neurotransmitter are still only partially understood. A similar hypothesis may be tested, however, with another modulatory substance, the neuropeptide FMRF-amide. Like histamine, FMRF-amide has distinctive and well-characterized inhibitory effects on *Aplysia* neurons, particularly on the clusters of sensory cells located in the animal's abdominal and pleural ganglia. The neuropeptide hyperpolarizes the membrane potential of these neurons and consequently decreases neurotransmitter release by causing the opening of a subclass of potassium channels, called S-channels.[22]

To determine whether arachidonate metabolites participate in the effects of FMRF-amide, a series of biochemical, pharmacological and electrophysiological experiments were carried out.[23] It was found that exposing primary cultures of sensory neurons to exogenous FMRF-amide elicits the formation of 12-lipoxygenase and 5-lipoxygenase metabolites. It was also observed that pharmacological inhibitors of phospholipase and lipoxygenase activities prevent the electrophysiological responses of sensory neurons to the application of FMRF-amide, whereas cyclooxygenase inhibitors have no effect. Finally, applications of arachidonic acid or of the 12-lipoxygenase metabolite, 12-HPETE, mimic the effects of FMRF-amide on potassium S-channels and on neurotransmitter release. By contrast, a variety of other lipoxygenase products (12-HETE, 5-HPETE and 5-HPETE) are ineffective.[23]

12-HPETE is a primary product of 12-lipoxygenase metabolism and, as we have seen in chapter 4, it can rapidly undergo a series of further metabolic transformations. Several such transformations have been shown to occur enzymatically in *Aplysia* neural tissue and to be stimulated by neuronal activity[24,25] (Fig. 5.4). Importantly, at least two of the 12-lipoxygenase products isolated from *Aplysia* neural tissue, hepoxilin A_3 and 12-ketoeicosatetraenoic acid (12-KETE), have also been shown to be biologically active. For example, when hepoxilin A_3 is applied onto the histamine-sensitive neuron, L14, it produces a dual action response (fast depolarization followed by slow hyperpolarization) similar to that caused by histamine.[24]

To determine whether 12-HPETE metabolism participates in the FMRF-amide response, Francesco Belardetti and his collaborators (at the University of Texas Southwestern Medical Center in Dallas) carried out a series of electrophysiological experiments using the technique of patch-clamping.[25] They confirmed that exogenous 12-HPETE activates potassium S-channels in intact *Aplysia* sensory neurons, but they also discovered that this hydroperoxide is inactive when applied to cell-free patches of neuronal membranes. The ability of 12-HPETE to activate the S-channels can be completely restored by incubating the membrane patches in the presence of hematin, which forms, by chemical catalysis, an array of 12-HPETE products qualitatively similar to those synthesized by the neurons (hepoxilins, 12-HETE and 12-KETE, Fig. 5.4). Moreover, the compound SKF-525-A, an inhibitor of cytochrome P_{450} activity, reduces the responses elicited by FMRF-amide, arachidonic acid and 12-HPETE in intact cells. These results suggest that a heme-containing enzyme, such as cytochrome P_{450}, may be responsible for converting 12-HPETE into an active metabolite (possibly hepoxilin A_3) which may then stimulate potassium S-channel opening by acting directly on the channel molecule or on some regulatory element associated with the channel (Fig. 5.5).[25]

Like those of other modulatory neuropeptides, the electrophysiological effects of FMRF-amide occur within seconds of its application and last in general as long as the application lasts. Prolonged exposure to FMRF-amide may result, however, in much longer-lasting changes in neuronal excitability. Piergiorgio Montarolo, Samuel Schacher and their colleagues at Columbia University found that exposing primary cultures of *Aplysia* sensory neurons to FMRF-

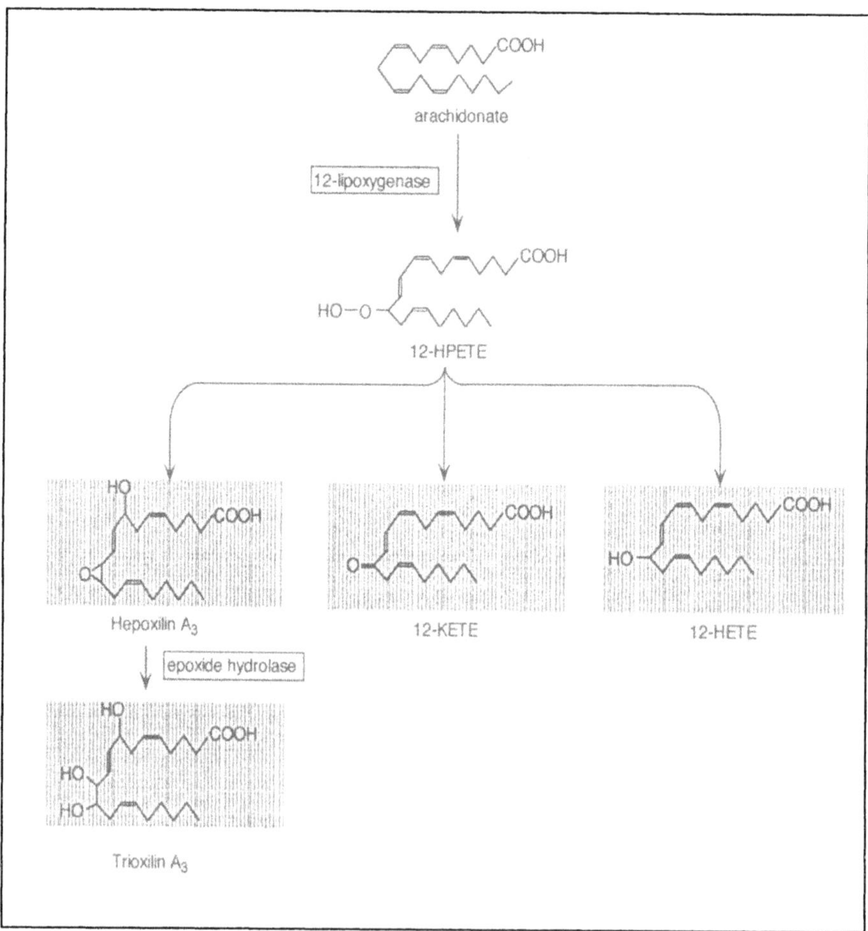

Fig. 5.4. 12-Lipoxygenase metabolism in the nervous system of the marine mollusk, Aplysia californica. *Highlighted are products that have been identified after physiological stimulations of nervous tissue. Abbreviations used: 12-hydro(pero)xyeicosatetraenoic acid, 12-H(P)ETE; 12-ketoeicosatetraenoic acid, 12-KETE. Based on results published in refs. 21, 24 and 25.*

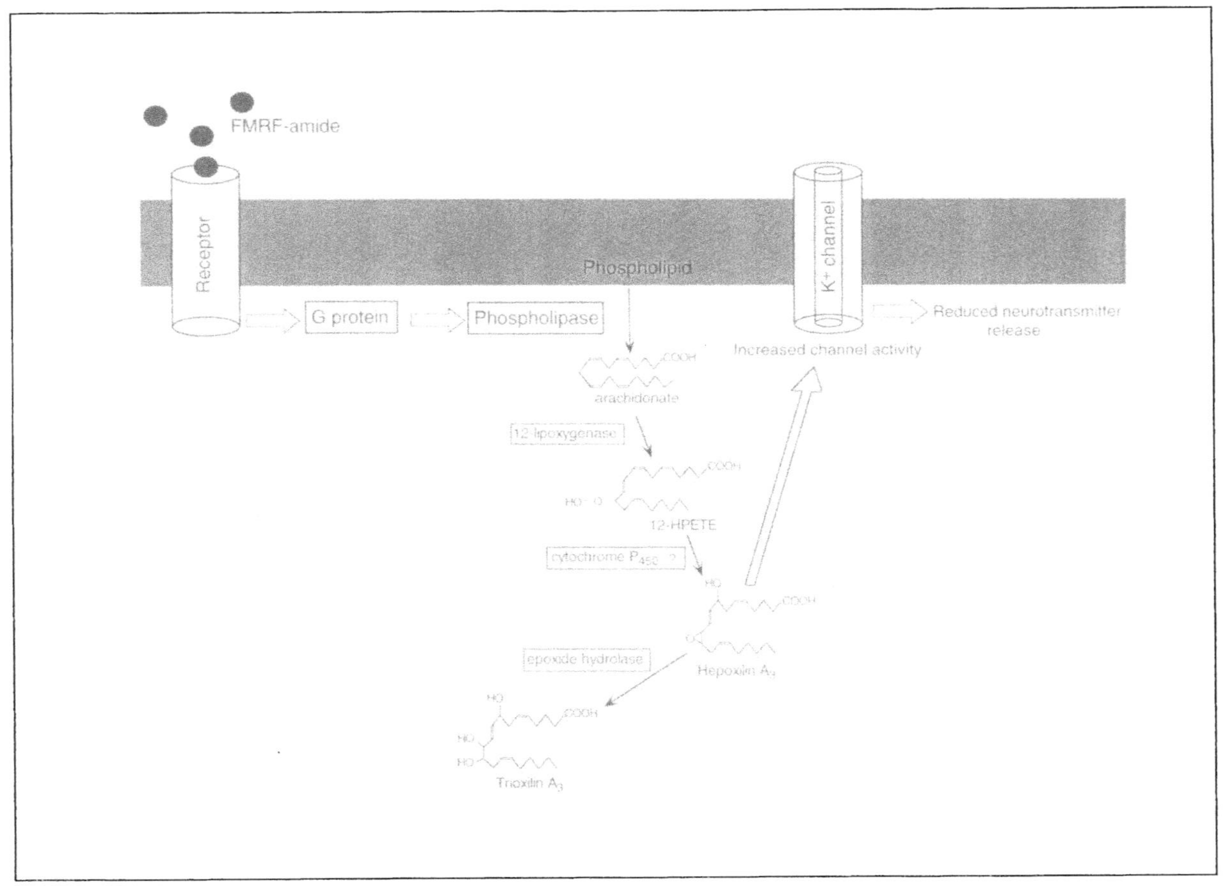

Fig. 5.5. Model of the second messenger mechanisms implicated in potassium channel modulation by lipoxygenase metabolites of arachidonic acid in Aplysia *neurons. Stimulation of FMRF-amide receptors may trigger phospholipase activation and arachidonate mobilization from membrane phospholipids. Non-esterified arachidonate may be metabolized to its primary 12-lipoxygenase product, 12-hydroperoxyeicosatetraenoic acid (12-HPETE), which may be then converted into hepoxilin A_3 and other products. One such product, possibly hepoxilin A_3 itself, may enhance potassium channel activity directly, by binding to the channel or to a regulatory element associated with it. Increased potassium channel activity would then lead to membrane hyperpolarization, and to decreased neurotransmitter release.*

amide for 2 hours causes a depression in synaptic transmission that can persist for several days. This synaptic depression is accompanied by a marked reduction in the number of transmitter-containing varicosities on neuronal dendrites.[26] As with the short-term modulatory effect discussed above, this form of long-term depression appears to be mediated by lipoxygenase metabolites of arachidonic acid. In support of this conclusion, it was found that exogenous arachidonate simulates the long-term response to FMRF-amide, and that this response can be prevented by treatment with lipoxygenase inhibitors.[27-29]

Effects reminiscent of those described in *Aplysia* have been reported in vertebrate nervous systems. For example, a role for lipoxygenase products in modulating potassium M-type current has been suggested both in pyramidal neurons from the rat hippocampal formation and in neurons from the bullfrog sympathetic ganglia.[30-33] Moreover, in the rat hippocampus, 12-lipoxygenase products were also shown to inhibit the transient calcium rise and the release of glutamate elicited by membrane depolarization.[34]

Regulation of Calcium/Calmodulin-Dependent Protein Kinase II by 12-Lipoxygenase Products

Along with ion channel activity, protein phosphorylation plays a fundamental role in regulating the release of neurotransmitters from synaptic terminals. A protein selectively associated with vertebrate nerve terminals, synapsin I, may be particularly important in this respect. Strong experimental evidence indicates, in fact, that the state of phosphorylation of this protein participates in regulating the number of neurotransmitter-containing vesicles that are immediately available for exocytosis.[35]

In its dephosphorylated state, synapsin I is thought to cross-link synaptic vesicles to the surrounding cytoskeletal lattice. When synapsin I is phosphorylated its interactions both with the synaptic vesicles and the cytoskeletal elements are reduced, resulting in a dissociation of the vesicles from the cytoskeleton. As a result of this phosphorylation reaction, which is catalyzed by the calcium/calmodulin protein kinase, CaM-kinase II, more vesicles may become available for fusion with the nerve terminal membrane and, consequently, a greater quantity of neurotransmitter may be released during nerve activity.[35]

12-Lipoxygenase metabolites may participate in regulating neurotransmitter release partly through a mechanism involving the inhibition of CaM-kinase II and the reduction of synapsin I phosphorylation. Experiments carried out in Paul Greengard's laboratory at the Rockefeller University showed that lipoxygenase-derived eicosanoids are potent in inhibiting purified CaM-kinase II activity. For example, 12-HPETE inhibits the kinase activity with a half-maximal effect at a concentration of 0.7 μM. By contrast, 12-HPETE has no effect on the activities of several other protein kinases (protein kinase C, protein kinase A, CaM-kinase I and CaM-kinase III) or on the protein phosphatase, calcineurin.[36]

The dual effects of 12-lipoxygenase metabolites on potassium channel and protein kinase activities have been integrated in a hypothetical model of their potential role in presynaptic inhibition. According to this model, illustrated schematically in Figure 5.6, synthesis of 12-HPETE within the nerve terminal may cause activation of potassium channels resulting in decreased calcium entry. In addition, 12-HPETE may cause inhibition of CaM-kinase II activity resulting in decreased phosphorylation of synapsin I and other synaptic phosphoproteins. These effects are expected to be synergistic in reducing transmitter release and synaptic efficacy.[37]

HOW DO EICOSANOIDS EXIT CELLS?

Arachidonate is mobilized in the cell's interior, and all eicosanoid-synthesizing enzymes are intracellular. Yet, aside from second messenger roles, such as those illustrated in the previous sections, all other actions of the eicosanoids take place in the extracellular milieu, where these molecules can interact with transmembrane receptors. How do the eicosanoids exit the cells in which they are produced?

It is often assumed that lipid molecules can readily traverse cell membranes by passive diffusion and freely access all cellular compartments without the need for any transport system. Despite its persistent popularity, this simplistic view is not likely to be correct. Strongly ionized lipids, such as phosphatidylcholine, are expected, of course, to require transport proteins because their charge would limit membrane diffusion. Indeed, ATP-dependent translocases that selectively extrude phospholipids from cells have

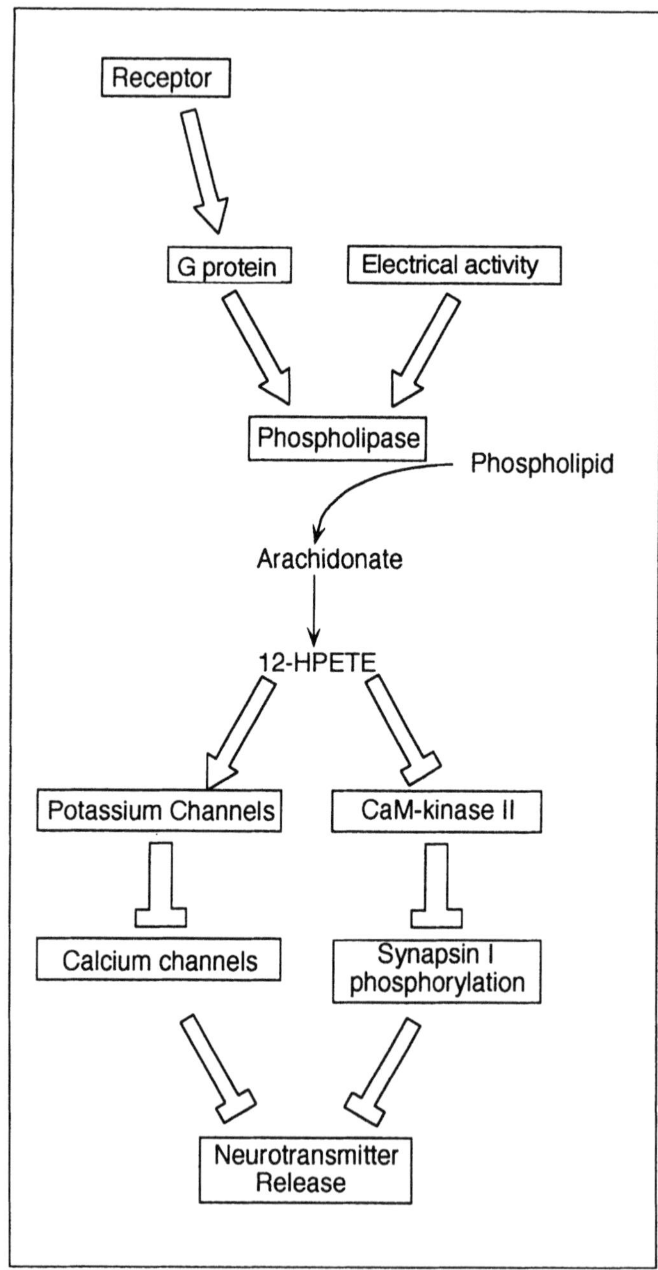

Fig. 5.6. Schematic model of the hypothetical roles of 12-HPETE in presynaptic inhibition of neurotransmitter release. 12-HPETE may both activate potassium channels, leading to decreased calcium entry, and inhibit CaM-kinase II, leading to decreased phosphorylation of synapsin I and other synaptic phosphoproteins. Together, these effects are expected to be synergistic in reducing neurotransmitter release.

been identified.[38] But even weakly charged lipid molecules, such as eicosanoids and other fatty acids, have been shown to require membrane carriers.

A number of studies over the past 10 years have described the kinetic properties and protease sensitivities of cellular fatty acid transfer systems. A general conclusion of these studies is that fatty acid transfer cannot be accounted for simply by a process of passive diffusion.[39-43] Moreover, at least three structurally different plasma membrane proteins that may mediate fatty acid transport have been identified by molecular cloning and expression in heterologous systems. These are the fatty acid transporter (FAT),[44] the fatty acid transport protein (FATP)[45] and the plasma membrane fatty acid-binding protein ($FABP_{pm}$).[46]

Even before an unequivocal demonstration of carrier-mediated fatty acid transport was provided, the pioneering studies of Laszo Bito and his coworkers at Columbia University had suggested the existence of a transport system for the prostaglandins. Prostaglandin accumulation against a concentration gradient was shown to take place in a variety of tissues, including the choroid plexus, the iris-ciliary body complex and the kidney cortex. Moreover, the prostaglandin transport was found to be saturable, energy-dependent and sensitive to certain chemicals, such as bromcresol green.[47-50]

These pharmacological studies have recently found strong support in the molecular cloning of a gene encoding for a prostaglandin transporter (PGT) protein enriched in organs containing epithelia. When expressed in tumor cells, the PGT protein confers to these cells the ability to take up certain prostaglandins (e.g., PGE_2 and $PGF_{2\alpha}$), but not others (e.g., TXB_2, 6-keto-$PGF_{1\alpha}$). PGT-induced prostaglandin uptake is sodium- and pH-insensitive, and can be inhibited by bromcresol green.[51]

What may be the physiological roles of the PGT protein? Likely possibilities include the transport of prostaglandins across epithelia and the uptake of circulating prostaglandins for intracellular metabolism, as suggested by Bito's original experiments. From the standpoint of the present discussion, most pertinent may be PGT's role in exporting newly synthesized prostaglandins into the extracellular space.

The discovery of PGT is quite recent, and many questions still lie ahead. We don't know yet if PGT is the only protein dedicated to prostanoid transport in mammalian tissues. Neither do we know of selective drugs that can block its activity. These issues are relevant, not only to our understanding of prostanoid biology, but also, in light of the multiple roles of the prostaglandins in disease, to the development of novel classes of therapeutic agents.

The existence of export carriers for the leukotrienes has also been postulated, mainly on the grounds of the hydrophilic properties of these compounds. LTC_4, for example, contains a glutathione moiety that would presumably hinder the molecule's diffusion through cell membranes.

To determine whether LTB_4 is actively extruded form cells, Takao Shimizu and his collaborators at the University of Tokyo have injected $[^3H]LTB_4$ into *Xenopus* oocytes, while measuring the efflux of radioactivity into the bathing solution. Their results support the existence in these oocytes of a selective transport system for LTB_4.[52] An unexpected protein, the multidrug resistance-associated protein (MRP), has been identified as a possible export carrier for LTC_4. The MRP, which is encoded by a gene overexpressed in various tumor cells that are resistant to anti-neoplastic drugs, transports LTC_4 through the cell membrane in an ATP-dependent manner. Its favorable kinetic properties ($K_M = 97$ nM) and substrate specificity suggest that it may mediate the extrusion of LTC_4 from stimulated cells.[53] Although the transport of both LTB_4 and LTC_4 have been shown to require energy, they appear to be markedly different in their sensitivities to pharmacological agents, suggesting the participation of distinct molecular entities.

THE PROSTANOID RECEPTORS

The idea that extracellular prostanoids exert their biological actions by activating selective membrane receptors is based on three lines of pharmacological evidence (for review, see refs. 54 and 55). Functional studies show that low concentrations of prostanoids elicit distinct sets of responses in different tissues. In some cases, such responses can be selectively mimicked by synthetic prostanoid agonists, or prevented by prostanoid antagonists. Studies with radioactively labeled ligands lend support to these observations: specific, high-affinity binding sites for the prostanoids can be readily demonstrated in membranes prepared from tissues that display functional responses to the prostanoids. Furthermore, these responses are always accompanied by the formation of intracellular second messengers, as it is expected of hormones or drugs that activate transmembrane receptors.

These pharmacological findings have now been supplemented by a great deal of structural information issued from the molecular cloning and characterization of prostanoid receptors. The first achievement in this direction stemmed from the chemical synthesis of potent TXA_2 receptor antagonists, such as the compound S-145, which was used as a tool to purify by affinity chromatography the TXA_2 receptor from human platelets.[56] Partial amino acid sequencing of the purified receptor allowed, in turn, the cloning and structural characterization of its complementary DNA.[57]

Thanks to this *tour de force*, accomplished in the laboratory of Shuh Narumiya in Kyoto, all known prostanoid receptors—a total of eight, without counting alternative splicing variants—could be subsequently cloned by homology screening.[58-69] Some of the common structural features of these receptors are illustrated schematically in Figure 5.7, and a synopsis of their pharmacological profiles, distribution in tissues and biological effects is provided in Table 5.1. It appears that most prostanoids activate only one receptor subtype, with the notable exception of PGE_2. PGE_2, endowed with different structures and pharmacological properties, activates at least four subtypes.

Like other G protein-coupled receptors, prostanoid receptors contain seven hydrophobic segments that probably transverse several times the plasma membrane, delimiting a series of extracellular and intracellular segments and loops (Fig. 5.7). Although the prostanoid receptors may have derived from a common ancestor, they are encoded by distinct genes, and their overall sequence homologies are relatively low. The most prominent regions of homology among them—but not with other G-protein-coupled receptor subtypes—are shown in Figure 5.7. These regions, limited to two segments found in the seventh transmembrane domain and in the second extracellular loop of the receptor, are thought to participate in ligand recognition.[70]

Table 5.1

Receptor subtype	Selective agonist	Selective antagonist	Tissue Distribution	Biological effects	Transduction system
DP	BW245C	BWA868C	platelets, smooth muscle, CNS	smooth muscle relaxation, inhib. platelet aggregation, sleep	↑cAMP
EP_1	Iloprost (also an IP agonist)	AH6809	smooth muscle	smooth muscle contraction	↑$[Calcium]_i$
EP_2	Butaprost	None	smooth muscle, leukocytes,mast cells, sensory neurons, lung, kidney	smooth muscle relaxation, inhib. mast cell secretion, stim. intestinal secretion	↑cAMP
EP_3	Enprostil	None	smooth muscle, adipocytes,CNS, sensory neurons, lung, kidney, endothelium, etc.	smooth muscle contraction, inhib. lipolysis and water reabsorption in kidney, stomach acid secretion	↓cAMP ↑$[Calcium]_i$
EP_4	None	AH22921 (also a TP receptor antagonist)	saphenous vein	relaxation	↑cAMP
FP	Fluprostenol	None	smooth muscle, corpus luteum, kidney	smooth muscle contraction, luteolysis	↑$[Calcium]_i$
IP	Cicaprost	None	platelets, smooth muscle, sensory neurons	smooth muscle relaxation, inhib. platelet aggregation	↑cAMP
TP	U46619	AH23848	platelets, smooth muscle, kidney, endothelium	smooth muscle contraction, stimul. platelet aggregation	↑$[Calcium]_i$

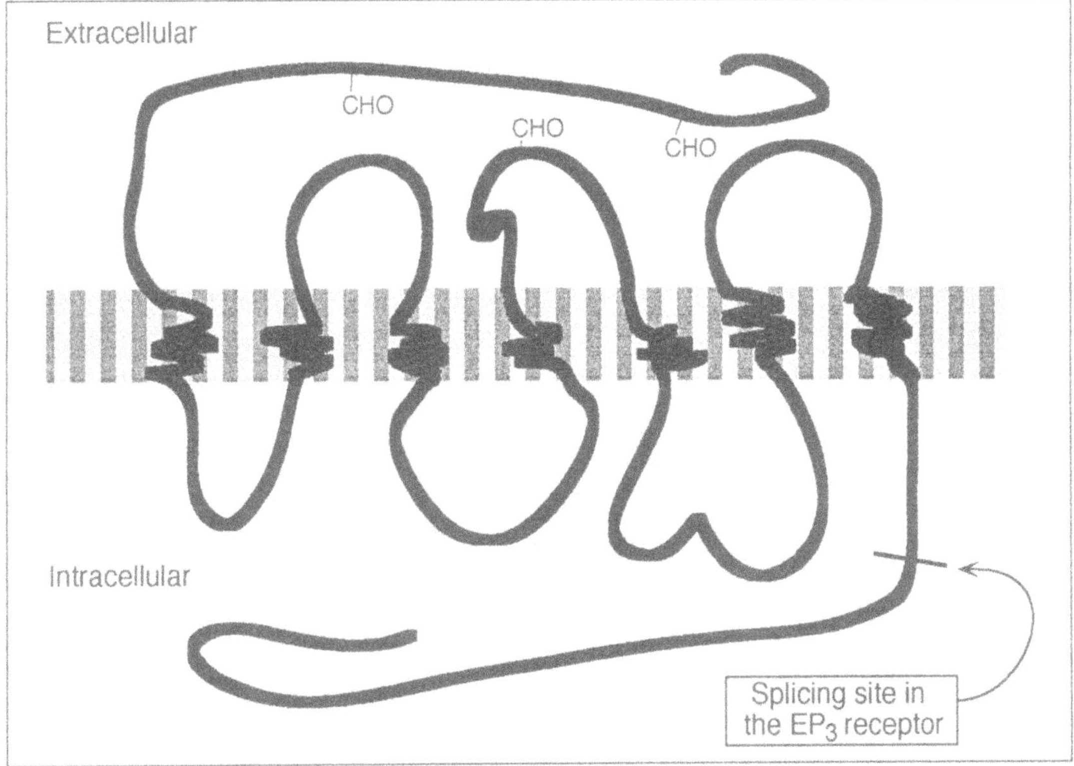

Fig. 5.7. General structure of prostanoid receptors. Like other members of the G protein-linked receptor superfamily, prostanoid receptors may span the plasma membrane seven times, forming a series of extracellular and intracellular domains. Highlighted, are regions that show the highest degree of sequence homology within the prostanoid receptor family. These regions are thought to participate in ligand binding. Potential sites of protein glycosylation are also indicated (CHO).

Signaling at Prostanoid Receptors

Another way of looking at the relations among prostanoid receptors is from the standpoint of the signal transduction mechanisms that they engage. Figure 5.8 depicts a phylogenetic tree obtained by comparing the deduced amino acid sequences of the cloned human prostanoid receptors.[69] Such comparison delineates two main receptor subfamilies—one coupled to the stimulation of adenylate cyclase activity (DP, IP, EP_2) and another to the increase in intracellular calcium levels (TP, EP_1, FP, EP_3).

As a rule, prostanoid receptors that are coupled to adenylate cyclase activation do so via a Gs protein, and they can be thought of as inhibitory. Thus, in the mammalian vascular apparatus, IP, DP and EP_2 receptors invariably mediate relaxation of smooth muscle cells in the vessel wall and inhibition of platelet aggregation (Table 5.1 and Fig. 5.9). Exceptions to this general principle are known to exist, though. For instance, in some species of teleost fish, PGI_2 is a potent vasoconstricting agent.[71] Moreover, in mammalian sensory neurons, the effects of PGI_2 are markedly

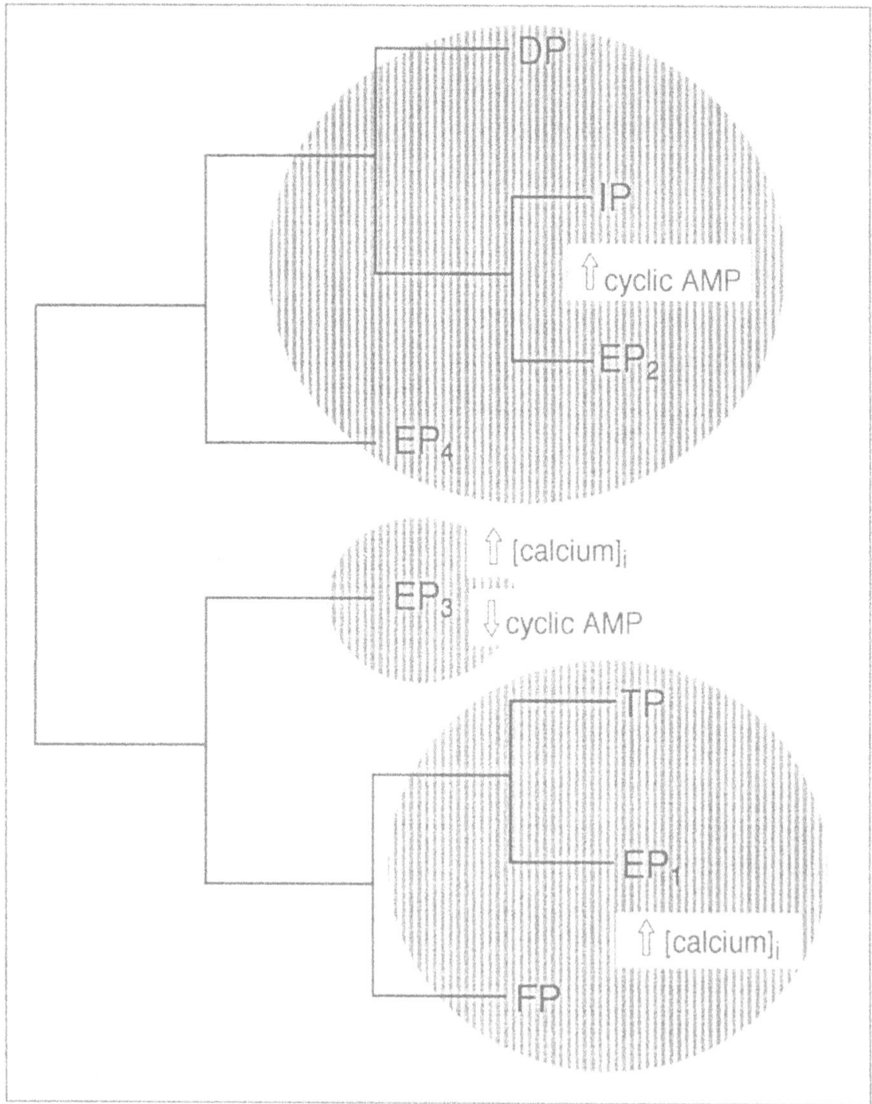

Fig. 5.8. Phylogenetic comparison of the prostanoid receptor family. Sequence analysis allowed definition of two main subfamilies of prostanoid receptors, which may also be distinguished by their different signal transduction mechanisms. (Redrawn from ref. 69.)

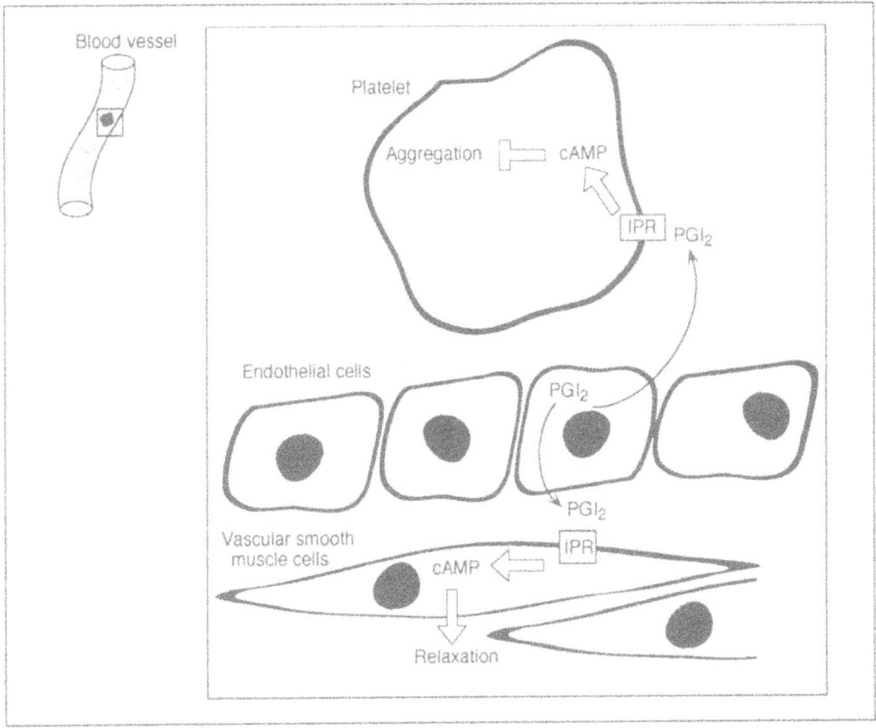

Fig. 5.9. Effects of vascular PGI_2 receptors, a prototype of adenylate cyclase-linked prostanoid receptors. PGI_2 released from endothelial cells interacts with IP receptors (IPR) present on vascular smooth muscle cells and blood platelets. IP receptors stimulate adenylate cyclase activity, which results in vasorelaxation and inhibition of platelet aggregation.

excitatory, suggesting a participation of this prostanoid in pain and hyperalgesia (see ref. 72 for review).

The coupling of prostanoid receptors to calcium increments has, ordinarily, excitatory physiological consequences. This effect cannot be ascribed to a single biochemical mechanism, however. For example, TP receptors in platelets mobilize calcium from intracellular stores through the sequential activation of a Gq protein and of a phosphoinositide-specific phospholipase C, resulting in inositoltriphosphate formation. A similar mechanism is invoked in vascular smooth muscle, where TP receptors mediate the potent vasoconstricting effects of TXA_2 (Fig. 5.10) (for review, see ref. 73). By contrast, EP_1 receptors can increase calcium levels either by mobilizing calcium from intracellular stores[74] or by opening receptor-

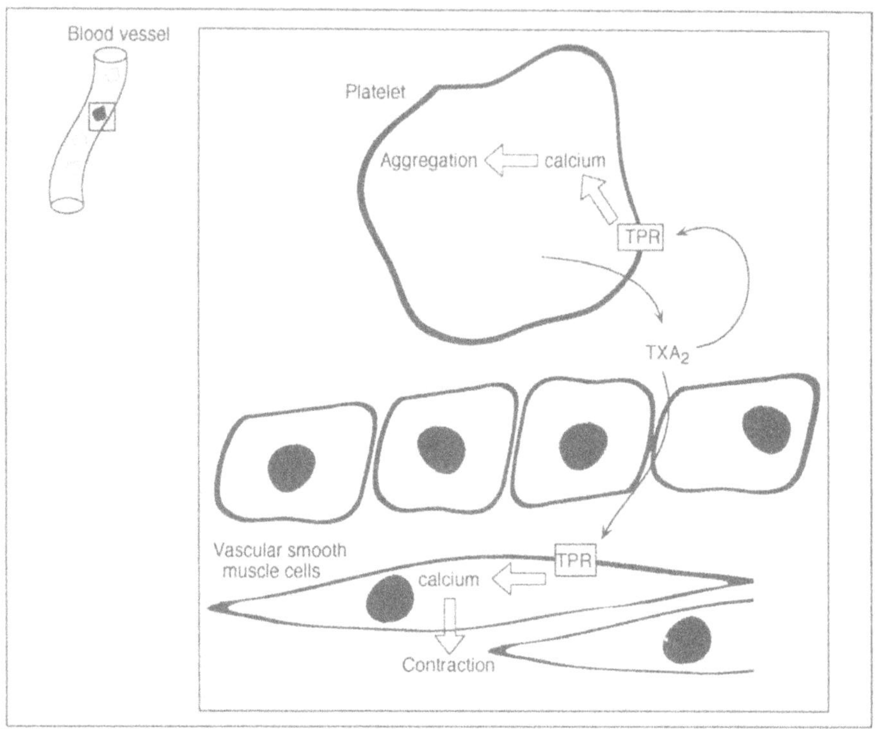

Fig. 5.10. Effects of vascular TXA_2 receptors, a prototype of prostanoid receptors linked to calcium mobilization. TXA_2 released from platelets activates TP receptors (TPR) on blood platelets and vascular smooth muscle. TP receptors stimulate phospholipase C activity, leading to inositoltrisphosphate accumulation and calcium mobilization from intracellular stores. This series of biochemical events elicits, in turn, contraction of vascular smooth muscle and platelet aggregation.

operated membrane channels that permit calcium entry from the extracellular space.[75]

Within the subfamily of prostanoid receptors that are linked to calcium increments, the EP_3 subtype stands out for a notable feature: the promiscuity of its coupling to intracellular second messengers. Like other members of its subfamily, the EP_3 receptor can activate a Gq protein and elicit calcium rises, which are thought to underlie biological responses such as smooth muscle contraction. But the EP_3 receptors can be also linked to the inhibition of adenylate cyclase activity. This effect, mediated via Gi, may be responsible for the reduction of acid secretion produced by PGE_2 in gastric mucosa. Finally, in some tissues, EP_3 receptor stimulation may also result in the activation of adenylate cyclase.[54]

The three signaling mechanisms engaged by EP_3 receptors may be coordinated at the level of mRNA transcription. Shuh Narumiya and his collaborators have cloned four alternative mRNA splicing variants of the EP_3 receptor from a bovine adrenal medulla cDNA library. The sequences of these variants are virtually identical, except for a stretch of amino acids constituting the carboxyl terminal tails, which have different lengths in each of the four isoforms (Fig. 5.7). That molecular differences in the carboxyl terminus may play an essential role in signaling is indicated by the fact that the four variants are coupled to distinct G proteins and second messenger systems, although their binding affinities for PGE_2 are very similar. The EP_3A isoform is linked to Gi/Go (inhibits adenylate cyclase), the EP_3D isoform to Gq (stimulates intracellular calcium mobilization), and the EP_3B and C isoforms to Gs (stimulate adenylate cyclase).[76] Underscoring the general relevance of this mechanism, a similar palette of splicing and signaling variants has been recognized in mouse, human and rabbit tissues.[63,77-78]

THE LEUKOTRIENE RECEPTORS

The potent biological effects exerted by the leukotrienes in vertebrate tissues are mediated by two groups of high-affinity membrane receptors, the LTB_4 receptor (BLT), and the peptidoleukotriene (cysteinylleukotriene) receptors ($CysLT_1$ and $CysLT_2$) (for review, see refs. 79 and 80). The pharmacological properties and tissue distributions of these receptor subtypes are illustrated in Table 5.2. While BLT receptors show a high selectivity for LTB_4, the two CysLT subtypes do not distinguish very well among the various peptidoleukotrienes (LTC_4, LTD_4 and LTE_4). Thus, although $CysLT_1$ shows some preference for LTC_4, and $CysLT_2$ for LTD_4, these selectivities are not absolute.

In many cells, the stimulation of $CysLT_1$ receptors is positively linked to inositol triphosphate turnover via a G protein (Gi/Go or Gq) and can therefore elicit calcium rises as well as arachidonate mobilization from membrane phospholipids.[81-83] A similar signaling mechanism has been demonstrated for the BLT receptor.[84-86] These cellular effects are likely to participate in some physiological processes. For instance, LTC_4 may regulate acetylcholine-activated

Table 5.2.

Receptor subtype	Selective agonist	Selective antagonist	Tissue Distribution	Biological effects	Transduction system
BLT	LTB_4	LY255283	neutrophils, smooth muscle,	chemotaxis, chemokinesis, smooth muscle contraction	$\uparrow$[Calcium]$_i$
$cysLT_1$	None	ICI1988615	smooth muscle, pulmonary epithelia, eosinophils	smooth muscle contraction, mucus secretion, eosinophil migration	$\uparrow$[Calcium]$_i$
$cysLT_2$	None	None	smooth muscle	smooth muscle contraction	$\uparrow$[Calcium]$_i$?

potassium currents in myocytes from guinea pig and bullfrog atrial tissue.[87,88] When unchecked, however, these effects may also have profound pathological repercussions.

A well-studied example in this regard is the involvement of the peptidoleukotrienes in the pathogenesis of bronchial asthma. Leukotrienes are released from human lung tissue by antigene challenge (see chapter 4), and they elicit an array of functional effects that tally with the pathological changes occurring in asthmatic lung tissue. These include increased microvascular permeability (which leads to edema formation), eosinophil influx to the inflamed areas, enhanced mucus secretion, bronchoconstriction, proliferation of airway smooth muscle cells and sensitization of sensory nerve fibers (for review, see refs. 89 and 90). A schematic view of the manifold effects of peptidoleukotrienes in asthma is provided in Figure 5.11. Preventing them, either by reducing leukotriene biosynthesis or by antagonizing their receptor-mediated response, will likely provide an effective approach to asthma therapy (for review, see refs. 91-92).

Because none of the leukotriene receptors has been cloned yet, the only information available on their molecular structures comes from biochemical studies. These have mainly relied on the techniques of photoaffinity labeling and radiation inactivation. Photolabeling of a human monocytic cell line with $[^3H]LTB_4$ has made it possible to determine the molecular mass of BLT receptors (50-60 kDa) and to establish the coupling of this receptor to G protein activation.[93] Using a similar approach, an LTD_4 receptor protein of 45 kDa was also identified in guinea pig lungs.[94]

A HIGH-AFFINITY LIPOXIN RECEPTOR

The lipoxins are produced by the sequential actions of multiple lipoxygenases on arachidonic acid and exert a variety of biological effects. In human neutrophils, the responses to lipoxin A_4 include inhibition of chemotaxis and calcium mobilization, and they are likely to be mediated by a high-affinity membrane receptor coupled to a pertussis toxin-sensitive G protein (Gi or Go) (see chapter 4 and ref. 95).

A complementary DNA encoding for what may be a lipoxin A_4 receptor was identified in the laboratory of Charles N. Serhan.[96]

This was an orphan complementary DNA, i.e., it had been previously cloned by sequence homology with other G protein-coupled receptors, but its ligand had not been identified. When it was expressed in Chinese hamster ovary cells, this DNA conferred to cells the ability to bind lipoxin A_4 with high affinity and to respond to the compound in a pertussis toxin-sensitive manner. Among the eicosanoids tested, only LTD_4 displaced lipoxin A_4 binding with high affinity (with a $K_i = 80$ nM). Whether this receptor binds, in vivo, exclusively to lipoxin A_4 or also to LTD_4, remains to be established.

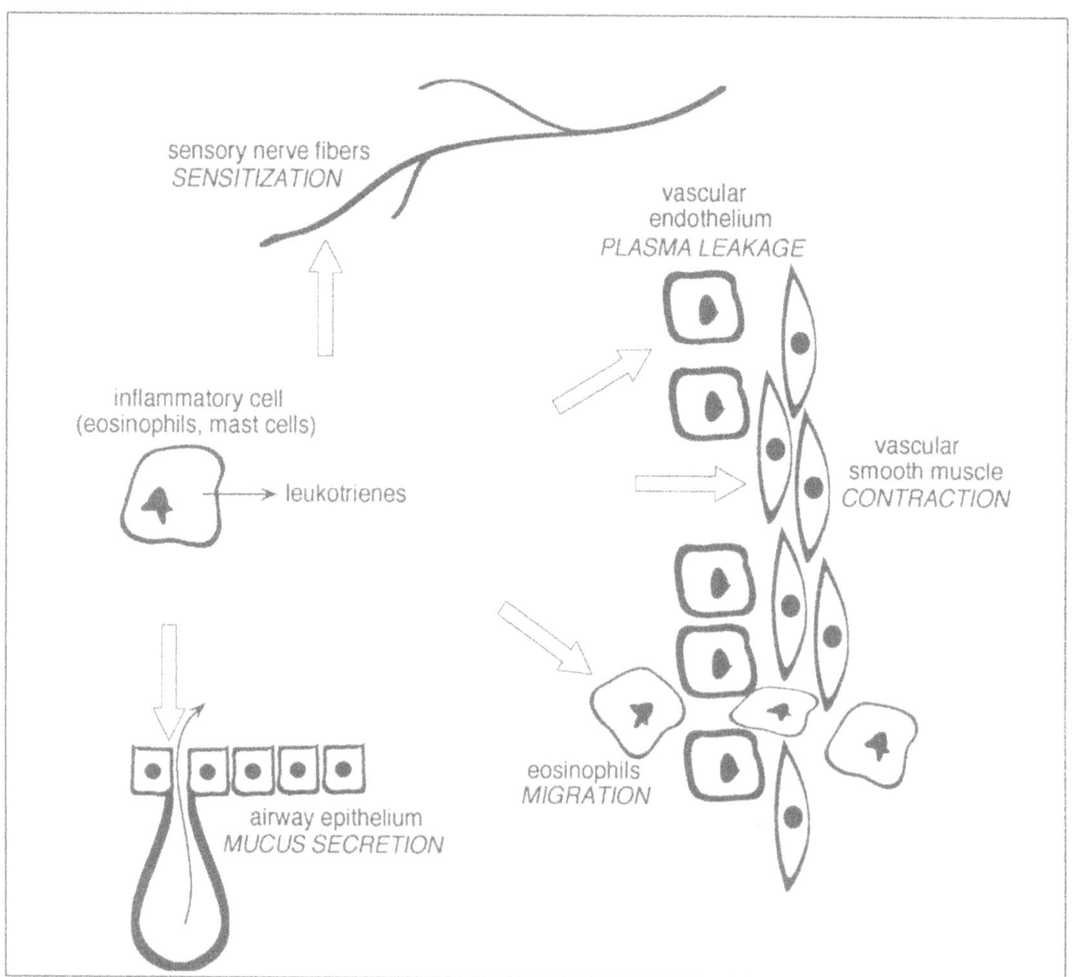

Fig. 5.11. Possible roles of the peptidoleukotrienes (LTC_4, LTD_4, LTE_4) in the pathogenesis of asthma.

REFERENCES

1. Nestler EJ, Greengard P. Protein Phosphorylation in the Nervous System. New York:Wiley, 1984.
2. O'Brian CA, Ward NE, Weinstein IB, Bull AW, Marnett LJ. Activation of rat brain protein kinase C by lipid oxidation products. Biochem Biophys Res Commun 1988; 155:1374-1380.
3. Hansson A, Serhan CN, Haeggström J, Ingelman-Sundberg M, Samuelsson B. Activation of protein kinase C by lipoxin A and other eicosanoids. Intracellular action of oxygenation products of arachidonic acid. Biochem Biophys Res Commun 1986; 134:1215-1222.
4. McGiff JC. Cytochrome P_{450} metabolism of arachidonic acid. Annu Rev Pharmacol Toxicol 1991; 31:339-369.
5. Hu S, Kim HS. Activation of K^+ channel in vascular smooth muscle by cytochrome P_{450} metabolites of arachidonic acid. Eur J Pharmacol 1993; 230:215-221.
6. Yu SP. Roles of arachidonic acid, lipoxygenases and phosphatases in calcium-dependent modulation of M-current in bullfrog sympathetic neurons. J Physiol (London) 1995; 487:797-811.
7. Taussig R, Sweet-Cordero A, Scheller RH. Modulation of ionic currents in *Aplysia* motor neuron B15 by serotonin, neuropeptides, and second messengers. J Neurosci 1989; 9:3218-3229.
8. Piomelli D. Arachidonic acid in cell signaling. Current Opinion Cell Biology 1993; 5:274-280.
9. Piomelli D. Arachidonic acid. In: Bloom FE, Kupfer DJ, eds. Psychopharmacology: The Fourth Generation of Progress. New York, Raven Press 1995; 595-607.
10. Evans RM. The steroid and thyroid hormone receptor superfamily. Science 1988; 240:889895.
11. Tontonoz P, Hu E, Spiegelman BM. Stimulation of adipogenesis in fibroblasts by PPARγ, a lipid-activated transcription factor. Cell 1994; 79:1147-1156.
12. Tontonoz P, Hu E, Graves RA, Budavari AI, Spiegelman BM. mPPARγ: tissue-specific regulator of an adipocyte enhancer. Genes Dev 1994; 8:1224-1234.
13. Kliever SA, Umesono K, Noonan DJ, Evans RM. Convergence of 9-retinoic acid and peroxisome proliferator signalling pathways through heterodimer formation of their receptors. Nature 1992; 358:771-774.
14. Göttlicher M, Widmark E, Li Q, Gustafsson J-Å. Fatty acids activate a chimera of the clofibric acid-activated receptor and the glucocorticoid receptor. Proc Natl Acad Sci USA 1992; 89:4563-4567.
15. Keller H, Deyer C, Medin J, Mahfoudi A, Ozato K, Wahli W. Fatty acids and retinoids control lipid metabolism through activation of peroxisome proliferator-activated receptor-retinoid X receptor heterodimers. Proc Natl Acad Sci USA 1993; 90:2160-2164.

16. Banner CD, Göttlicher M, Widmark E, Sjovall J, Rafter JJ, Gustafsson J-Å. A systematic analytical/cell assay approach to isolate activators of orphan nuclear receptors from biological extracts: characterization of peroxisome proliferator-activated receptor activators in plasma. J Lipid Res 1993; 34:1583-1591.
17. Forman BM, Tontonoz P, Chen J, Brun RP, Spiegelman BM, Evans RM. 15-deoxy-$\Delta^{12,14}$-prostaglandin J_2 is a ligand for the adipocyte determination factor PPARγ. Cell 1995; 83:803-812.
18. Kliewer SA, Lenhard JM, Willson TM, Patel I, Morris DC, Lehmann JM. A prostaglandin J_2 metabolite binds peroxisome proliferator-activated receptor γ and promotes adipocytes differentiation. Cell 1995; 83:813-819.
19. Narumiya S, Ohno K, Fukushima M, Fujiwara M. Site and mechanism of growth inhibition by prostaglandins. III. Distribution and binding of prostaglandin A_2 and Δ-^{12}-prostaglandin J_2 in nuclei. J Pharmacol Exp Therap 1987; 242:306-311.
20. Armstrong DL, White RE. An enzymatic mechanism for potassium channel stimulation through pertussis-toxin-sensitive G proteins. Trends Neurosci 1992; 15:403-408.
21. Piomelli D, Shapiro E, Feinmark SJ, Schwartz JH. Metabolites of arachidonic acid in the nervous system of *Aplysia*: possible mediators of synaptic modulation. J Neurosci 1987; 7:3675-3686.
22. Belardetti F, Kandel ER, Siegelbaum SA. Neuronal inhibition by the peptide FMRFamide involves opening of S-K^+ channels in *Aplysia*. Nature 1987; 325:153-156.
23. Piomelli D, Volterra A, Dale N, Siegelbaum SA, Kandel ER, Schwartz JH, Belardetti F. Lipoxygenase metabolites of arachidonic acid as second messengers for presynaptic inhibition of *Aplysia* sensory cells. Nature 1987; 328:38-43.
24. Piomelli D, Shapiro E, Zipkin R, Schwartz JH, Feinmark SJ. Formation and action of 8-hydroxy-11,12-epoxy-icosatrienoic acid in *Aplysia*: a possible second messenger in neurons. Proc Natl Acad Sci USA 1989; 86:1721-1725.
25. Piomelli D, Feinmark SJ, Schwartz JH. Formation and biological activity of 12-keto-eicosatetraenoic acid in the nervous system of *Aplysia*. J Biol Chem 1988; 263:16591-16596.
26. Montarolo PG, Kandel ER, Schacher S. Long-term heterosynaptic inhibition in *Aplysia*. Nature 1988; 333:171-174.
27. Schacher S, Kandel ER, Montarolo PG. cAMP and arachidonic acid simulate long-term structural and functional changes produced by neurotransmitters in Aplysia sensory neurons. Neuron 1993; 10:1079-1088.

28. Wu F, Schacher S. Pre- and postsynaptic changes mediated by two second messengers contribute to expression of Aplysia long term heterosynaptic inhibition. Neuron 1994; 12:407-421.
29. Byrne JH, Zwatijes R, Homayouni R, Critz SD, Eskin A. Roles of second messenger pathways in neuronal plasticity and in learning and memory. Insights gained from *Aplysia*. In: Shenolikar S, Nairn AC, eds. Advances Second Messengers Phosphoprotein. Res Vol. 27, New York: Raven Press, 1993; 47-108.
30. Schweitzer P, Madamba S, Siggins GR. Arachidonic acid metabolites as mediators of somatostatin-induced increase of neuronal M-current. Nature 1990; 346:464-467.
31. Schweitzer P, Madamba S, Champagnat J, Siggins GR. Somatostatin inhibition of hippocampal CA1 pyramidal neurons: mediation by arachidonic acid and its metabolites. J Neurosci 13:2033-2049.
32. Lammers CH, Schweitzer P, Facchinetti P, Arrang JM, Madamba S, Siggins GR, Piomelli D. Arachidonate 5-lipoxygenase and its activating protein: prominent hippocampal expression and role in somatostatin signaling. J Neurochem 1996; 66:147-152.
33. Yu SP. Roles of arachidonic acid, lipoxygenases and phosphatases in calcium-dependent modulation of M-current in bullfrog sympathetic neurons. J Physiol 1995; 487:797-811.
34. Freeman EJ, Damron DS, Terrian DM, Dorman RV. 12-lipoxygenase products attenuate the glutamate release and Ca^{2+} accumulation evoked by depolarization of hippocampal mossy fiber nerve endings. J Neurochem 1991; 56:1079-1082.
35. Greengard P, Valtorta F, Czernik AJ, Benfenati F. Synaptic vesicle phosphoproteins and regulation of synaptic function. Science 1993; 259:780-785.
36. Piomelli D, Wang JKT, Sihra TS, Nairn AC, Czernik AJ, Greengard P. Inhibition of Ca^{2+}/calmodulin-dependent protein kinase II by arachidonic acid metabolites. Proc Natl Acad Sci USA 1989; 86:8550-8554.
37. Piomelli D, Greengard P. Lipoxygenase metabolites of arachidonic acid in neuronal transmembrane signaling. Trends Pharmacol Sci 1990; 11:367-373.
38. Ruetz S, Gross P. Phosphatidylcholine translocase: a physiological role for the mdr2 gene. Cell 1994; 77:1071-1081.
39. Abumrad NA, Perkins RC, Park JH, Park CR. Mechanism of long chain fatty acid permeation in the isolated adipocyte. J Biol Chem 1981; 256:9183-9191.
40. Abumrad NA, Park JH, Park CR. Permeation of long chain fatty acid into adipocytes. Kinetics, specificity and evidence for involvement of a membrane protein. J Biol Chem 1984; 259:8945-8953.

41. Stremmel W, Berk PD. Hepatocellular influx of [^{14}C]oleate reflects membrane transport rather than intracellular metabolism or binding. Proc Natl Acad Sci USA 1986; 83:3086-3090.
42. Bojesen IN, Bojesen E. Arachidonic acid transfer across the human red cell membrane by a specific transport system. Acta Physiol Scand 1995; 154:253-267.
43. Bojesen IN, Bojesen E. Oleic acid binding and transport capacity of human red cell membrane. Acta Physiol Scand 1996; 156: 501-516.
44. Abumrad NA, el-Maghrabi MR, Amri EZ, Lopez E, Grimaldi PA. Cloning of a rat adipocyte membrane protein implicated in binding or transport of long chain fatty acids that is induced during preadipocyte differentiation. J Biol Chem 1993; 268:17665-17668.
45. Schaffer JE, Lodish HF. Expression cloning and characterization of a novel adipocyte long chain fatty acid transport protein. Cell 1994; 79:427-436.
46. Isola LM, Zhou S-L, Kiang C-L, Stump DD, Bradbury MW, Berk PD. 3T3 fibroblasts transfected with a cDNA for mitochondrial aspartate aminotransferase express plasma membrane fatty acid-binding protein and saturable fatty acid uptake. Proc Natl Acad Sci USA 1995; 92:9866-9870.
47. Bito LZ. Saturable, energy-dependent, transmembrane transport of prostaglandins against concentration gradients. Nature 1975; 256:134-136.
48. Bito LZ, Davson H, Hollingsworth JR. Facilitated transport of prostaglandins across the blood-cerebrospinal fluid and blood-brain barriers. J Physiol 1976; 256:273-285.
49. Bito LZ, Davson H, Salvador EV. Inhibiiton of in vitro concentrative prostaglandin accumulation by prostaglandins, prostaglandin analogues and by some inhibitors of organic anion transport. J Physiol 1976; 256:257-271.
50. DiBenedetto FE, Bito LZ. Transport of prostaglandins and other eicosanoids by the choroid plexus: its characterization and physiological significance. J Neurochem 1986; 46:1725-1731.
51. Knai N, Lu R, Satriano JA, Bao Y, Wolkoff AW, Schuster VL. Identification and characterization of a prostaglandin transporter. Science 1995; 268:866-869.
52. Mori M, Izumi T, Shimizu T. Energy-dependent export of leukotriene B_4 in *Xenopus* oocytes. Biochim Biophys Acta 1993; 1168:23-29.
53. Leier I, Jedlitschky G, Buchholz U, Cole SP, Deeley RG, Keppler D. The MRP gene encodes an ATP-dependent export pump for leukotriene C_4 and structurally related conjugates. J Biol Chem 1994; 269:27807-27810.

54. Coleman RA, Smith WL, Narumiya S. Classification of prostanoid receptors: properties, distribution and structure of the receptors and their subtypes. Pharmacol Rev 1994; 46:205-229.
55. Negishi M, Sugimoto Y, Ichikawa A. Prostanoid receptors and their biological actions. Prog Lip Res 1993; 32:417-434.
56. Ushikubi F, Nakajima M, Hirata M, Okuma M, Fujiwara M, Narumiya S. Purification of the thromboxane A_2/prostaglandin H_2 receptor from human blood platelets. J Biol Chem 1989; 264:16496-16501.
57. Hirata M, Hayashi Y, Ushikubi F, Yokota Y, Kageyama R, Nakanishi S, Narumiya S. Cloning and expression of cDNA for a human thromboxane A_2 receptor. Nature 1991; 349:617-620.
58. Hirata M, Kakizuka A, Aizawa M, Ushikubi F, Narumiya S. Molecular characterization of a mouse prostaglandin D receptor and functional expression of the cloned gene. Proc Natl Acad Sci USA 1994; 91:11192-11196.
59. Watabe A, Sugimoto Y, Honda A, Irie A, Namba T, Negishi M, Ito S, Narumiya S, and Ichikawa A. Cloning and expression of cDNA for a mouse EP_1 subtype of prostaglandin receptor. J Biol Chem 1993; 268:20175-20178.
60. Takeuchi K, Abe T, Takahashi N, Abe K. Molecular cloning and intrarenal localization of rat prostaglandin receptor EP_3 subtype. Biochem Biophys Res Commun 1993; 194:885-891.
61. Sugimoto Y, Namba T, Honda A, Hayashi Y, Negishi M, Ichikawa A, Narumiya S. Cloning and expression of a cDNA for mouse prostaglandin E receptor EP_3 subtype. J Biol Chem 1992; 267: 6463-6466.
62. Honda A, Sugimoto Y, Namba T, Watabe A, Irie A, Negishi M, Narumiya S, Ichikawa A. Cloning and expression of a cDNA for mouse prostaglandin E receptor EP_2 subtype. J Biol Chem 1993; 268:7759-7762.
63. Adam M, Boie Y, Rushmore TH, Müller G, Bastien L, KcKee KT, Mtters KM, Abramovitz M. Cloning and expression of the human EP_3 prostanoid receptor. FEBS Letters 1994; 338:170-174.
64. An S, Yang J, Xia M, Goetzl EJ. Cloning and expression of the EP_2 subtype of human receptors for prostaglandin E. Biochem Biophys Res Commun 1993; 197:263-270.
65. Sugimoto Y, Hasumoto K, Namba T, Irie A, Katsuyama M, Negishi M, Kakizuka A, Narumiya S, Ichikawa A. Cloning and expression of a cDNA for mouse prostaglandin F receptor. J Biol Chem 1994; 269:1356-1360.
66. Nüsing RM, Hirata M, Kakizuka A, Eki T, Ozawa K, Narumiya S. Characterization and chromosomal mapping of the human thromboxane A_2 receptor gene. J Biol Chem 1993 268:25253-25259.

67. Namba T, Oida H, Sugimoto Y, Kakizuka A, Negishi M, Ichikawa A, Narumiya S. cDNA cloning of a mouse prostacyclin receptor; multiple signaling pathways and expression in thymic medulla. J Biol Chem 1994; 269:9986-9992.
68. Namba T, Sugimoto Y, Hirata M, Hayashi Y, Honda A, Watabe A, Negishi M, Ichikawa A, Narumiya S. Mouse thromboxane A_2 receptor: cDNA cloning, expression and Northern blot analysis. Biochem Biophys Res Commun 1992; 184:1197-1203.
69. Boie Y, Sawyer N, Slipetz D, Metters KM, Abramovitz M. Molecular cloning and characterization of the human prostanoid DP receptor. J Biol Chem 1995; 270:18910-18916.
70. Narumiya S, Hirata M, Namba T, Hayashi Y, Ushikubi F, Sugimoto Y, Negishi M, Ichikawa A. Structure and function of prostanoid receptors. J Lip Mediators 1993; 6:155-161.
71. Piomelli D, Pinto A, Tota B. Divergence in vascular actions of prostacyclin during vertebrate evolution. J Experim Zool 1985; 233:127-131.
72. Wise H, Jones RL. Focus on prostacyclin and its novel mimetics. Trends Pharmacol Sci 1996; 17:17-21.
73. Armstrong RA, Wilson NH. Aspects of the thromboxane receptor system. Gen Pharmcol 1995; 26:463-472.
74. Funk CD, Furci L, Fitzgerald GA, Grygorczyk R, Rochette C, Bayne M, Abramovitz M, Adam M, Metters KM. Cloning and expression of a cDNA for the human prostaglandin E receptor EP1 subtype. J Biol Chem 1993; 268:26767-26772.
75. Creese RB, Denborough MA. The effects of prostaglandin E_2 on contractility and cyclic AMP levels of guinea-pig tracheal smooth muscle. Clin Exper Pharmacol Physiol 1981; 8:616-617.
76. Namba T, Sugimoto Y, Negishi M, Irie A, Ushikubi F, Kakizuka A, Ito S, Ichikawa A, Narumiya S. Alternative splicing of C-terminal tail of prostaglandin E receptor subtype EP3 determines G-protein specificity. Nature 1993; 365:166-170.
77. Irie A, Sugimoto Y, Namba T, Harazono A, Honda A, Watabe A, Negishi M, Narumiya S, Ichikawa A. Third isoform of the prostaglandin-E-receptor EP_3 subtype with different C-terminal tail coupling to both stimulation and inhibition of adenylate cyclase. Eur J Biochem 1993; 217:313-318.
78. Breyer MD, Jacobson HR, Davis LS, Breyer RM. In situ hybridisation and localisation of mRNA for the rabbit prostaglandin EP_3 receptor. Kidney Int 1993; 43:1372-1378.
79. Crooke ST, Mong S, Clark M, Hogaboom G, Lewis M, et al. Leukotriene receptors and signal transduction. Biochem Actions Horm 1987; 14:81-139.

80. Halushka PV, Mais DE, Mayeux PR, Morinelli TA. Thromboxane, prostaglandin and leukotriene receptors. Annu Rev Pharm Tox 1989; 10:213-239.
81. Sarau H, Mong S, Foley J, Wu H, Crooke ST. Identification and characterization of leukotriene D_4 receptors and signal transduction processes in rat basophilic leukemic cells. J Biol Chem 1987; 262:4034-4041.
82. Mong S, Wu H, Miller J, Hall R, Gleason J et al. SKF104353, a high affinity antagonist for human and guinea pig lung leukotriene D_4 receptor blocked phosphatidylinositol metabolism and thromboxane synthesis induced by leukotriene D_4. Mol Pharmacol 1986; 239:63-70.
83. Mong S, Miller J, Wu H, Clark M, Gleason J, Crooke ST. Leukotriene D_4 receptor-mediated synthesis and release of arachidonic acid metabolites in guinea pig lung: induction of thromboxane and prostacyclin biosynthesis by leukotriene D_4. J Pharmacol Exp Ther 1986; 239:63-70.
84. Goldman DW, Chang FH, GIfford LA, Goetzl EJ, Bourne HR. Pertussis toxin inhibition of chemotactic factor-induced calcium mobilization and function in human polymorphonuclear leukocytes. J Exp Med 1985; 162:145-156.
85. Andersson T, Schlegel W, Monod A, Krause KH, Stendhal O, Lew DP. Leukotriene B_4 stimulation of phagocytes results in the formation of inositol 1,4,5-triphosphate. A second messenger for Ca^{2+} mobilization. Biochem J 1986; 240:333-340.
86. Holian A. Leukotriene B_4 stimulation of phosphatidylinositol turnover in macrophages and inhibition by pertussis toxin. FEBS Lett 1986; 201:15-19.
87. Scherer RW, Lo CF, Breitweiser GE. Leukotriene C_4 modulation of muscarinic K^+ current activation in bullfrog atrial myocytes. J Gen Physiol 1993; 102:125-141.
88. Kurachi Y, Ito H, Sugimoto T, Shimizu T, Miki I, Ui M. α-Adrenergic activation of the muscarinic K^+ channel is mediated by arachidonic acid metabolites. Pflügers Arch 1989; 414:102-104.
89. Barnes PJ, Chung KF, Page CP. Inflammatory mediators and asthma. Pharmacol Rev 1988; 40:49-84.
90. Hay DWP, Torphy TJ, Undem BJ. Cysteinyl leukotrienes in asthma: old mediators up to new tricks. Trends Pharmacol Sci 1995; 16:304-309.
91. Chung KF. Leukotriene receptor antagonists and biosynthesis inhibitors: potential breakthrough in asthma therapy. Eur Resp J 1995; 8:1203-1213.
92. Snyder DW, Fleisch JH. Leukotriene receptor antagonists as potential therapeutic agents. Ann Rev Pharmacol Toxicol 1989; 29:123-143.

93. Slipetz DM, Scoggan KA, Nicholson DW, Metters KM. Photoaffinity labelling and radiation inactivation of the leukotriene B_4 receptor in human myeloid cells. Eur J Pharmacol 1993; 244:161-173.
94. Metters KM, Zamboni RJ. Photoaffinity labelling of the leukotriene C_4 receptor in guinea pig lung. J Biol Chem 1993; 268:6487-6495.
95. Serhan CN. Lipoxin biosynthesis and its impact in inflammatory and vascular events. Biochim Biophys Acta 1994; 1212:1-25.
96. Fiore S, Maddox JF, Perez HD, Serhan CN. Identification of a human cDNA encoding a functional high affinity lipoxin A_4 receptor. J Exp Med 1994; 180:253-260.

CHAPTER 6

ARACHIDONATE DERIVATIVES AS ENDOGENOUS CANNABINOID SUBSTANCES

Until 1992, most researchers working on the metabolism of arachidonic acid in the central nervous system were convinced that the brain mobilized and transformed this fatty acid basically as did all other tissues, producing the more or less expected array of bioactive eicosanoids which we have encountered in the preceding chapter. This simple view was radically challenged, however, by a report appeared in the December 18, 1992 issue of *Science* magazine. The report, originating in the laboratory of Raphael Mechoulam at the Hebrew University in Jerusalem, described the isolation and chemical identification of a bioactive substance present in brain, whose pharmacological properties are basically undistinguishable from those of Δ^9-tetrahydrocannabinol—the psychoactive principle of cannabis.[1]

This discovery was immediately hailed as a turning point in the neuropharmacology of *Cannabis*. A few years earlier, the finding of cannabinoid receptors expressed at high levels in discrete regions of the brain had already suggested that an endogenous cannabinoid substance might exist, but until Mechoulam's breakthrough, the structure of this hypothetical substance remained elusive. Now, researchers could begin to address important questions that had long remained intractable, such as whether endogenous cannabinoids play a role in controlling brain functions and mental states.

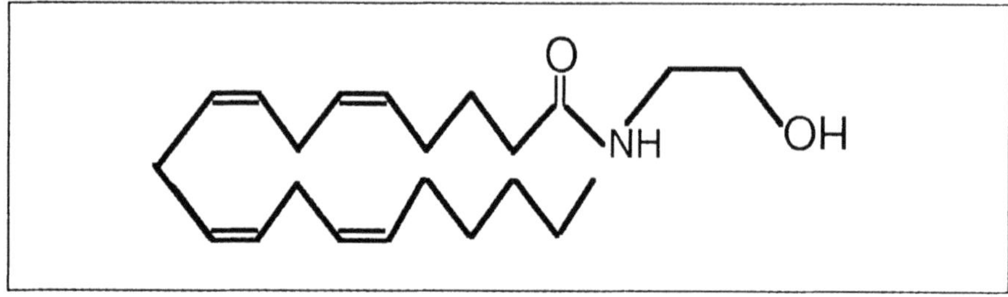

Fig. 6.1. The chemical structure of anandamide (N-arachidonoylethanolamine), an endogenous cannabinoid substance identified in mammalian brain tissue.

The interest in Mechoulam's study extended very rapidly from the field of *Cannabis* pharmacology to that of arachidonic acid biochemistry. In fact, it had not escaped the attention of many readers of the *Science* report that the chemical structure of the endogenous cannabinoid isolated by the Jerusalem's group—called *anandamide* after *ananda*, the Sanskrit word for 'bliss'—was that of a unique arachidonic acid derivative, N-arachidonoylethanolamine. This structure, depicted in Figure 6.1, clearly showed that anandamide was not produced through any of the then known metabolic transformations of free arachidonate.

What is, then, the biogenesis of anandamide? One of the aims of the present chapter is to review the last three years of research in this area and provide a possible answer to this question. I will also briefly describe what is currently known on other aspects of anandamide metabolism in brain tissue, in particular its hydrolytic degradation and oxygenation by lipoxygenase enzymes. Before we address these issues, however, it will be useful to review the biological effects of anandamide in the wider context of the multiple pharmacological properties of plant-derived and synthetic cannabinoid drugs.

THE PHARMACOLOGY OF CANNABIS

A detailed description of the pharmacological effects of *Cannabis* was already accessible to Greek and Roman physicians in the first century AD, when Dioscorides included the plant in his classic textbook of *Materia Medica* (Fig. 6.2). As expected from the wide distribution of *Cannabis* in the Far East, ancient Indian and Chinese medical writers were also familiar with its therapeutic usage and were indeed much more accurate than their European colleagues in describing the remarkable effects of the plant on the central nervous system (for review, see ref. 2 and 3).

We know now that these central effects—which, from animal and human studies, are often epitomized as a combination of euphoria, analgesia, hypomotility and hypothermia—are principally caused by the dibenzopyrane derivative, Δ^9-tetrahydrocannabinol,

Fig. 6.2. The Cannabis *plant, from a Byzantine manuscript of Dioscorides'* Materia Medica *(I century A.D.). Dioscorides was not aware of the psychotropic effects of* Cannabis, *but later medical writers, such as Oribasius (IV century A.D.), described them quite accurately.*

present in the yellow resin which covers the leaves and flower clusters of the ripe female plant.

The chemical structure of Δ^9-tetrahydrocannabinol, determined by Gaoni and Mechoulam in 1964,[4] is illustrated in Figure 6.3. Unlike many other biologically active chemicals of plant origin, Δ^9-tetrahydrocannabinol is a highly hydrophobic molecule, a property that has hindered the progress on its mode of action for nearly three decades. Indeed, not only was Δ^9-tetrahydrocannabinol more difficult to handle experimentally than such hydrophylic alkaloids as cocaine or morphine, but also its preference for lipid

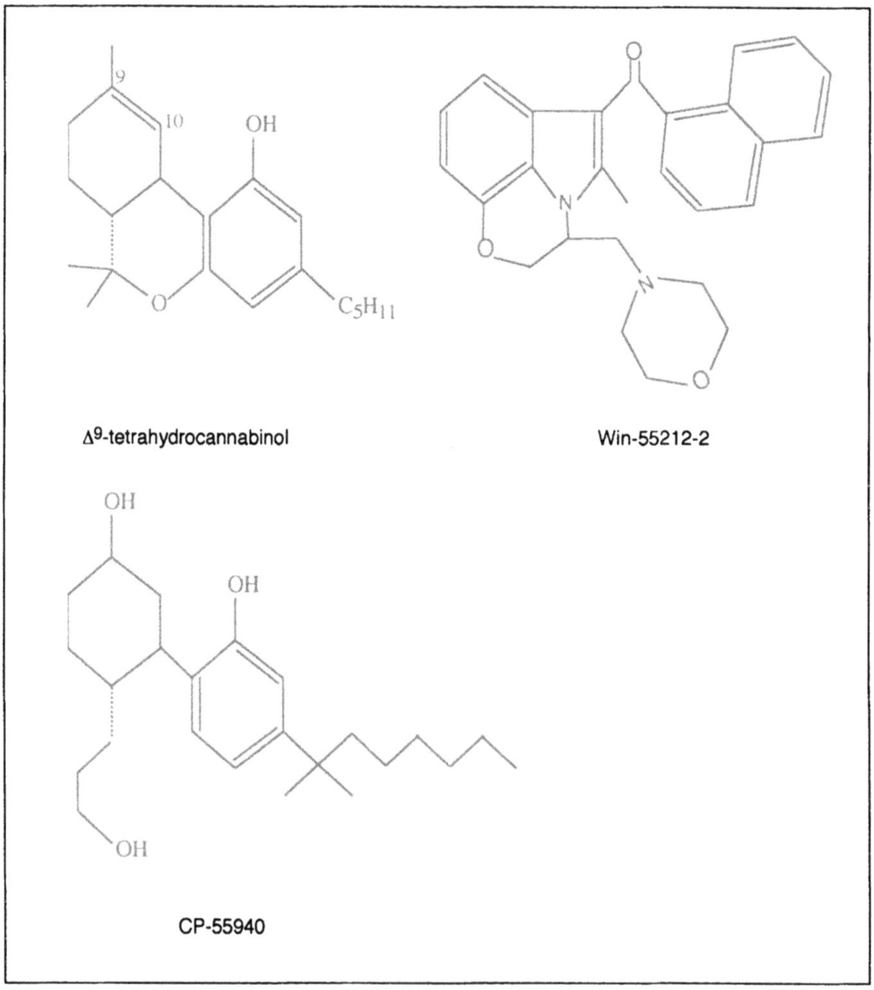

Δ^9-tetrahydrocannabinol

Win-55212-2

CP-55940

Fig. 6.3. Chemical structures of select cannabinoid receptor agonists.

membranes erroneously suggested that the drug's main effect was to modify the properties of cell membrane bilayers in a non-selective manner.[5]

Two events intervened to change this view. First, families of synthetic analogs of Δ^9-tetrahydrocannabinol were developed. These agents had pharmacological effects that were qualitatively identical to those of Δ^9-tetrahydrocannabinol but displayed greater potency and, most importantly, stereoselectivity (Fig. 6.3). Because stereoselectivity is irreconcilable with non-specific membrane interactions, the pharmacological profile of these drugs can be considered as the first evidence that selective cannabinoid receptors exist. Second, as a consequence of this development it became possible to reveal the existence of cannabinoid receptors by using standard pharmacological approaches, and high-affinity binding sites for cannabinoids were soon described in brain membrane preparations. Moreover, occupation of these binding sites was shown to be modulated by non-hydrolyzable analogs of guanosine triphosphate (GTP) and to be coupled to inhibition of adenylyl cyclase activity, providing evidence that they represented bona fide G protein-linked transmembrane receptors.[6,7] Conclusively supporting these observations, in 1990 the complementary DNA encoding a G protein-coupled cannabinoid receptor was cloned and characterized at the molecular level by Lisa Matsuda and collaborators at the National Institutes of Mental Health in Bethesda.[8]

In heterologous expression systems, the brain cannabinoid receptor (now referred to as CB1 receptor) is functionally coupled to multiple intracellular signaling pathways, including inhibition of adenylyl cyclase activity, inhibition of voltage-activated calcium channels and activation of a delayed rectifier potassium conductance (Fig. 6.4).[9-12] In situ hybridization studies demonstrated that CB1 receptor mRNA is abundantly expressed in discrete regions of the rat brain—including dorsal striatum, cerebral cortex, hippocampus, amygdala, hypothalamus and cerebellum—whereas it is expressed at very low levels in most non-nervous tissues.[13,14] Results consistent with the distribution of CB1 mRNA were obtained in autoradiographic localizations of the receptor protein. These experiments were carried out in the laboratory of Miles Herkenham at the National Institutes of Mental Health by using the potent cannabinoid receptor agonist, [^{3}H]CP-55,940.[15] The specific brain distribution of CB1 receptor protein and mRNA, demonstrated by these studies, provides an essential anatomical correlate for the cognitive and motor effects of cannabinoid drugs (Fig. 6.5).

In a sense, this distribution is not completely satisfying. For many years, in fact, researchers had known that the pharmacological effects of the cannabinoids are not restricted to the central nervous system. The multiple, potent actions of these agents on peripheral organs include reduction of pain and inflammation, lowering of intraocular pressure associated with glaucoma and relief of muscle spasms associated with multiple sclerosis.[16-18] How are all these effects brought about, if peripheral tissues contain exceedingly low levels of CB1 receptors?

A partial answer to this question, whose therapeutic significance should not be underestimated, came from the discovery of a second cannabinoid receptor restricted in its localization to macrophages resident in the marginal zone of spleen and, possibly, to peripheral mast cells.[19,25] This receptor, called CB2, has only about

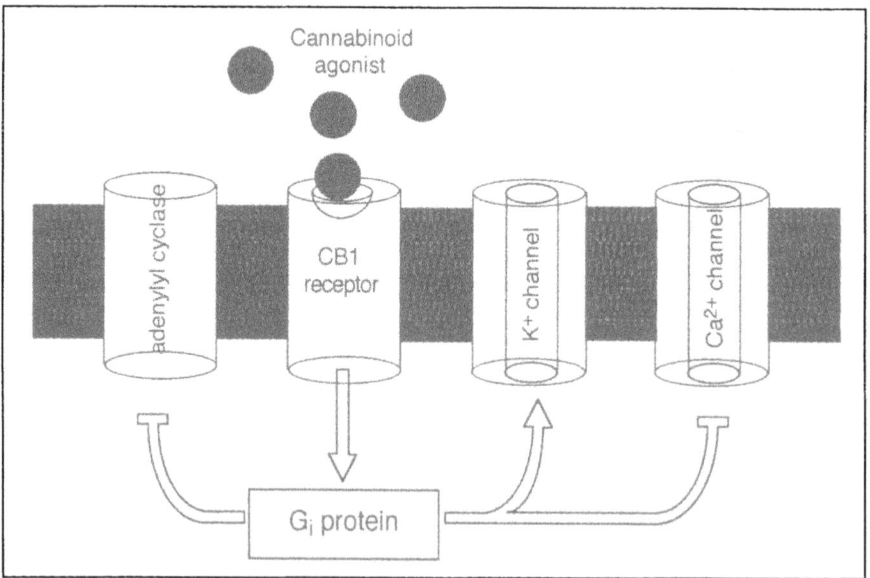

Fig. 6.4. Hypothetical model of the intracellular signaling pathways activated by cannabinoid CB_1 receptors. Stimulation of these receptors causes activation of potassium channels and inhibition of calcium channels and adenylate cyclase activity. The diagram combines results obtained in various heterologous expression systems, as described in refs. 9-12.

44% sequence identity with its brain counterpart, indicating that the two subtypes did not diverge recently in evolution (and that it should be possible therefore to design drugs able to discriminate between them). The intracellular coupling of the CB2 receptor has not yet been extensively investigated, but the available evidence suggests that it may resemble that of the CB1 receptor; for example, in transfected cells it also appears to be linked to the inhibition of adenylyl cyclase activity.[20]

SEARCHING FOR AN ENDOGENOUS CANNABINOID SUBSTANCE

Why do cannabinoid receptors exist? The experience with the enkephalins has taught neuropharmacologists that whenever receptors for a drug are present in the body, chances are that they are there to bind one or more endogenous factor(s). Not surprisingly, therefore, researchers have long been looking for chemicals, produced by the mammalian brain, that may activate cannabinoid receptors.

How does one go about such a task? One approach is based on the logical premise that an endogenous cannabinoid substance

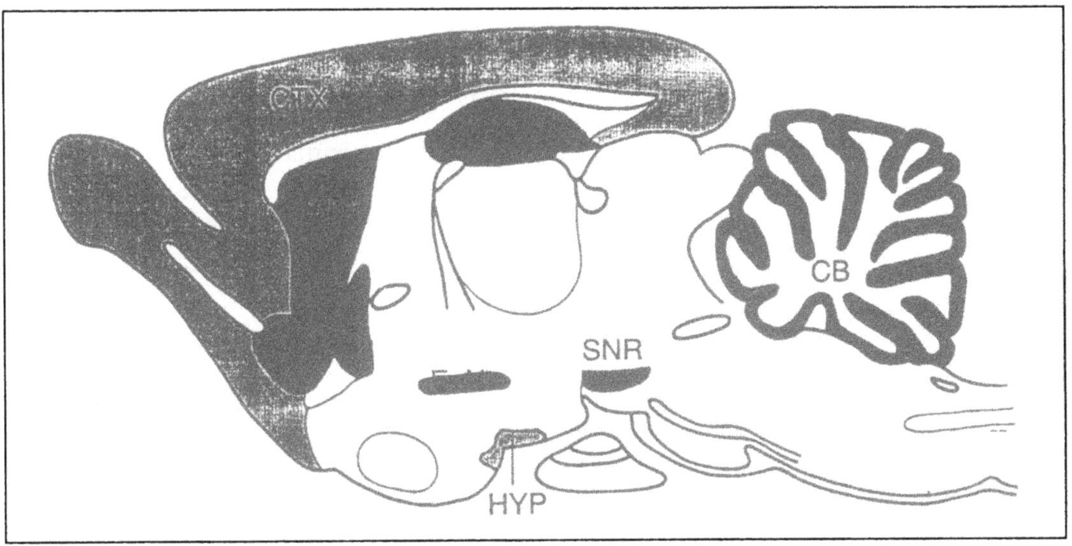

Fig. 6.5. Schematic distribution of cannabinoid CB1-type receptors in rat brain. Abbreviations used are: CB, cerebellum; CTX, cerebral cortex; EnN, entopenduncular nucleus; GP, globus pallidus; HIP, hippocampus; HYP, hypothalamus; SNR, substantia nigra pars reticulata; STR, striatum.

should have the properties of a neurotransmitter and, thus, should be released from brain tissue when this is appropriately stimulated. This approach, depicted schematically in Figure 6.6A, was adopted by researchers in Allyn Howlett's laboratory at the Saint Louis University School of Medicine. They incubated rat brain slices in the presence of a calcium ionophore and determined whether the media of these incubations contained a chemical factor capable of displacing the binding of radiolabeled CP-55940, a cannabinoid agonist, to brain membranes. While they could demonstrate that a cannabinoid-like binding activity was released from the stimulated slices, the insufficient quantities of material provided by their approach prevented a complete chemical characterization of this activity.[21,22]

An altogether different path was taken by the laboratory of Raphael Mechoulam—where, as we have just seen, Δ^9-tetrahydrocannabinol had been originally isolated. Raphael Mechoulam, William Devane and their coworkers hypothesized that hydrophobicity may be a chemical feature which cannabinoid drugs could have in common with the endogenous cannabinoid factor(s). So, instead of starting from conditioned media, they subjected whole porcine brains to organic solvent extraction and fractionated the lipid extract by using various chromatographic techniques: throughout the fractionation, they followed the fractions' ability to displace binding of a radiolabeled synthetic cannabinoid to brain membranes (Fig. 6.6B). From these fractions, they purified a lipid component, which they characterized by gas chromatography/mass spectrometry and by nuclear magnetic resonance spectroscopy as the ethanolamide of arachidonic acid—anandamide.[1,23] The mass spectrum of synthetic anandamide, obtained by electron-impact ionization, in shown in Figure 6.7.

The chemical synthesis of anandamide confirmed this structural identification, and allowed Mechoulam, Devane and their colleagues to determine its pharmacological properties. In vitro and in vivo tests showed a great similarity of effects between anandamide and cannabinoid drugs. Anandamide reduced the electrogenic contraction of mouse *vas deferens* and, most importantly, closely mimicked the behavioral responses induced by

A. ***Hypothesis tested***
Endogenous cannabinoid substances are released from neurons into the extracellular space by an activity-dependent process.

Brain slices, maintained *in vitro*, were stimulated with a calcium ionophore

The incubation medium was collected.

Samples of the incubation medium were tested for their ability to displace the binding of a labeled cannabinoid receptor agonist to brain membranes

Results
Calcium-dependent release of a cannabimimetic activity was found, but this activity was not identified chemically.

B. ***Hypothesis tested***
Endogenous cannabinoid substances are, like Δ^9-THC, hydrophobic molecules.

Whole brains were extracted with organic solvents

The organic extract was fractionated by chromatography.

The lipid fractions were tested for their ability to displace binding of a labeled cannabinoid receptor agonist to brain membranes

The bioactive lipid component was characterized by GC/MS and NMR

Results
The presence in brain of a cannabimimetic substance was established, but this was not identified as a neural modulatory substance.

Fig. 6.6. Experimental approaches taken by the laboratories of A. Howlett (A) and R. Mechoulam (B) in the search for an endogenous cannabinoid substance.

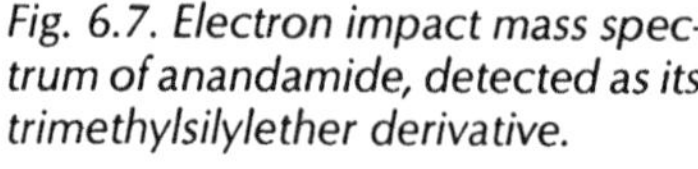

Fig. 6.7. Electron impact mass spectrum of anandamide, detected as its trimethylsilylether derivative.

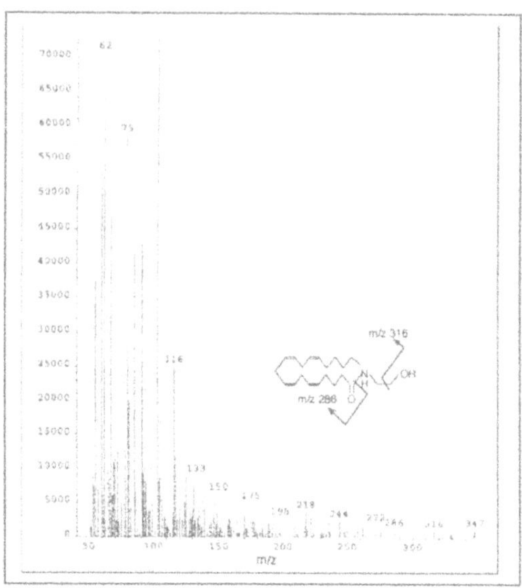

Δ^9-tetrahydrocannabinol in vivo: in the rat, the lipid was found to produce analgesia, hypothermia and hypomotility.[1,24]

Subsequently, the pharmacological properties of anandamide have been explored in considerable detail and by using a variety of experimental models, all of which have confirmed and extended the validity of the early observations. There is now little doubt that anandamide is an endogenous lipid component present in the brain, which activates CB1 cannabinoid receptors. Although anandamide binds also to peripheral CB2 receptors, it does so with a much lower affinity than to CB1. Moreover, its ability to *activate* the CB2 subtype remains controversial, and a role as endogenous CB2 receptor *antagonist* has been proposed.[25]

The biological functions of anandamide may extend beyond its ability to activate cannabinoid receptors. Recent evidence indicates that this arachidonate derivative may stimulate receptors that are pharmacologically distinct from known cannabinoid receptor subtypes. When applied to brain striatal astrocytes in primary culture, anandamide has a potent inhibitory effect on gap-junction conductance. As expected from a receptor-mediated event, this effect is specific for anandamide (e.g., N-eicosatrienoyl-ethanolamine is inactive and non-esterified arachidonate is 10-fold less active), and is prevented by treating the astrocytes with either pertussis toxin or N-ethylmaleimide (two agents that block the activation of G_i/G_o proteins). The fact that two cannabinoid receptor agonists do not mimic the inhibitory effect of anandamide, and that the cannabinoid antagonist, SR141716A, does not prevent it, indicates that this putative astrocyte receptor is pharmacologically different from either CB1 or CB2.[26]

We do not know yet the physiological significance of this response, but one intriguing possibility is that, by inhibiting gap-junction permeability, anandamide released from neurons may participate in controlling intercellular communication in astrocytes. Brain astrocytes are thought to communicate with one another through various non-synaptic mechanisms: one such mechanism, known as regenerative calcium waves, consists of the gap junction-mediated intercellular spread of calcium transients throughout an astrocyte syncytium.[27,28] Glutamate released from synaptic terminals evokes calcium waves in adjacent astrocytes. The waves propagate, in turn, throughout the astrocytic network and trigger responses in neurons embedded within it. Because calcium waves are mediated by gap junctions, which are blocked by anandamide, it is not surprising that anandamide also inhibits the calcium waves, raising the possibility that this lipid mediator may turn off calcium-mediated long-range signaling between neurons and astrocytes (Fig. 6.8).

THE REDISCOVERY OF THE N-ACYLETHANOLAMINES

The pharmacological similarities with Δ^9-tetrahydrocannabinol point to anandamide as a candidate for the role of endogenous cannabinoid substance. But they do not prove it. Proof that an endogenous chemical acts as a signaling molecule in the central nervous system must meet stringent criteria, which are not dissimilar from those for intracellular second messengers listed at the beginning of chapter 3. Essentially, it must be shown that brain cells are capable to produce, release and dispose of the bioactive chemical in a functionally significant manner—for instance, as seen with the 'classical' neurotransmitters glutamate and γ-aminobutyrate.

To determine whether anandamide meets the criteria of a neurotransmitter, my collaborators and I carried out a series of biochemical experiments using rat brain neurons in primary culture. We labeled the neurons in a medium containing radioactive ethanolamine and stimulated them with a calcium ionophore or with

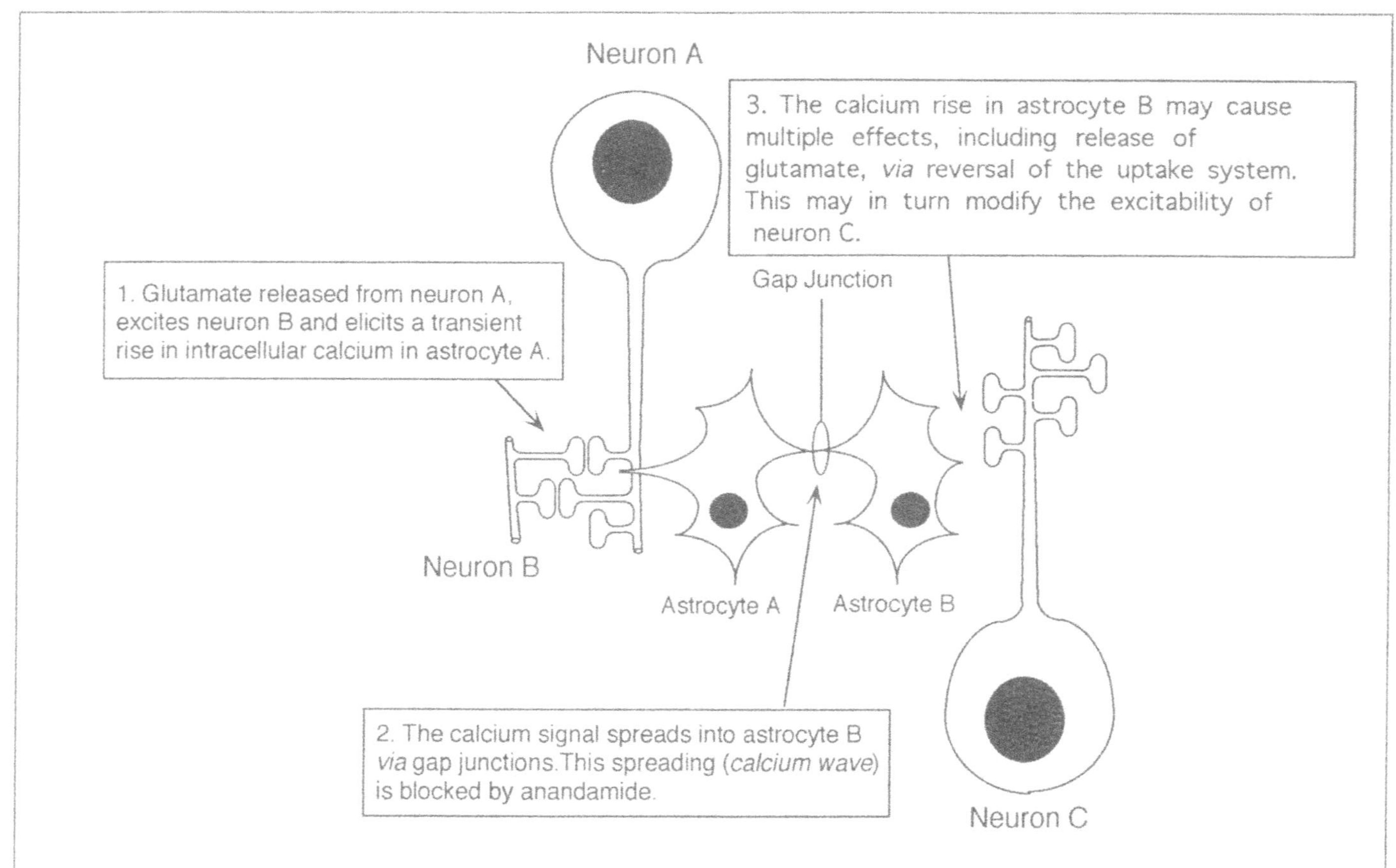

Fig. 6.8. Possible roles of gap junction-mediated calcium waves in brain astrocytes.

membrane depolarizing agents (e.g., the glutamate receptor agonist, kainate). We found that stimulated neurons produce radioactive anandamide, and that this production is both dependent on calcium ions and restricted to neurons, as one would expect of a neural signaling substance.[29] We were struck, however, by an unexpected finding: together with anandamide, a group of at least four additional radioactive components were also released from the neurons. By using HPLC and GC/MS, we identified these components as a family of N-acylethanolamines (NAEs) similar to anandamide in their general structure but differing in the fatty acyl chain bound to the ethanolamine moiety (Fig. 6.9).

It turns out that our surprise at the finding of multiple NAEs was largely unwarranted. Actually, for many years it had been known that brain tissue contains such compounds: they had been isolated first in 1963 by the laboratory of Sidney Udenfriend at the NIH[30] and subsequently studied by Harald H.O. Schmid and collaborators at the Hormel Institute of the University of Minnesota (for review, see ref. 31). In those studies no evidence had been found, however, for the natural occurrence of polyunsaturated NAEs (such as anandamide), probably because of technical difficulties in isolating minute amounts of these relatively unstable molecules. Technical problems may also explain why early studies did not uncover the stimulus-dependent generation of NAEs by neurons, which lead researchers in this area to conclude that NAE formation only occurred under conditions of conspicuous tissue damage, such as ischemia.[31]

The fact that saturated and monosaturated NAEs were co-released with anandamide from stimulated neurons raised the possibility that the former may also serve a physiological function related to cannabimimetic signaling.[29] This hypothesis finds support in a recent report. A major NAE released from neurons, N-palmitoylethanolamine, was found to bind to and activate a CB2-like receptor subtype expressed by peripheral mast cells and by the mast cell-derived cell line, RBL-2H3.[25] If confirmed, these observations would support the idea that a single biochemical event leads to the co-release from neurons of a family of cannabimimetic NAEs, acting on both CB1 and CB2 receptors. Let us examine now in greater detail the possible nature of this biochemical event.

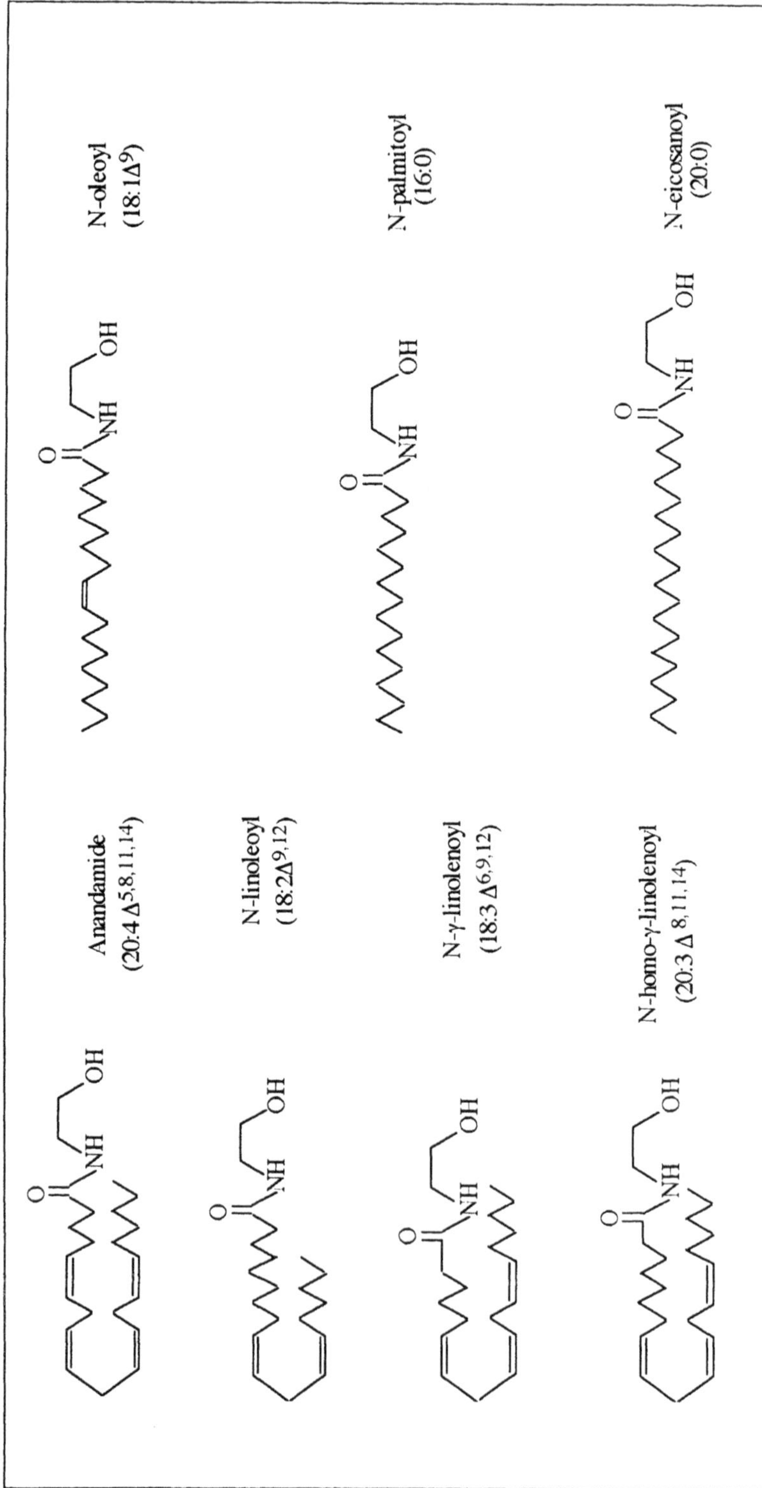

Fig. 6.9. Chemical structures of various N-acylethanolamines identified in mammalian brain tissue.[29,44]

FORMATION OF ANANDAMIDE AND OTHER N-ACYLETHANOLAMINES

Before the discovery of anandamide, many aspects of the biogenesis of NAEs had been unraveled in a series of experiments carried out in the laboratory of Harald H.O. Schmid.[31] These studies indicated that formation of NAEs likely proceeds from the hydrolytic cleavage of a phospholipid precursor, N-acylphosphatidylethanolamine (NAPE), rather than from the condensation reaction of fatty acids with ethanolamine which had been suggested by earlier studies (ref. 32 and see below). Our own work on primary cultures of rat brain neurons corroborated and extended the pathway predicted by Schmid and coworkers (Fig. 6.10).

Primary cultures of neurons, but not astrocytes, contain detectable quantities of a lipid component which we have identified as NAPE by enzymatic cleavage, multiple chromatographic analyses and nuclear magnetic resonance spectroscopy.[29] Neuronal NAPE is composed of a variety of molecular species, which differ in the fatty acyl group bound, through an amide bond, to the ethanolamine moiety of phosphatidylethanolamine (PE). We have found at least five such molecular species in cultured neurons (Table 6.1).

Table 6.1. N-acylethanolamines produced by primary cultures of rat striatal neurons[34]

	(% of total)
N-arachidonoylethanolamine	0.2
N-g-linolenoylethanolamine	5.0
N-linoleoylethanolamine	0.2
N-oleoylethanolamine	27
N-stearoylethanolamine	20
N-palmitoylethanolamine	52

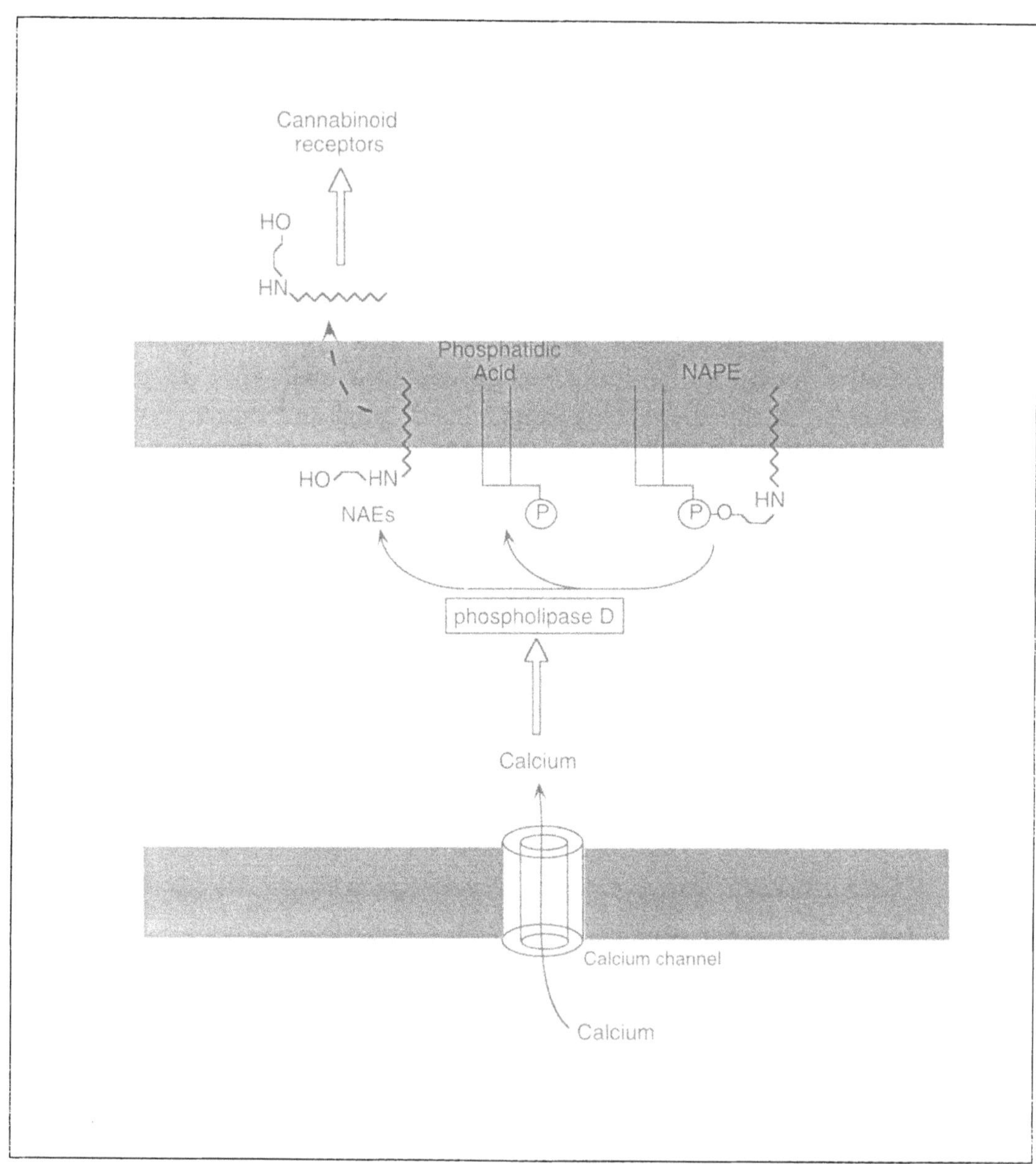

Fig. 6.10. Hypothetical scheme illustrating the possible mechanism of formation of anandamide and other N-acylethanolamines (NAEs) in primary cultures of rat brain neurons. Elevation of intracellular calcium levels results in the activation of a phospholipase D activity, which cleaves N-acylphosphatidylethanolamine (NAPE), forming anandamide and other NAEs. These lipids may then exit the neuron, through an as yet unknown transport system, and activate cannabinoid receptors located on neighboring cells.

How is NAPE synthesized, and how is its biosynthesis controlled? Unfortunately, we cannot satisfactorily answer this question at present. A calcium-dependent enzyme activity that catalyzes the transfer of fatty acyl groups from various donor phospholipids to PE, producing NAPE, has been partially characterized in dog brain tissue and in primary cultures of rat brain neurons.[33a,b] In these neurons, NAPE biosynthesis is stimulated by stimuli that elevate intracellular calcium levels and is potentiated by cAMP-dependent protein kinase activity.[33b] But the physiological roles, anatomical distribution and regulation of NAPE biosynthesis in the adult nervous system remain unknown.

In what cellular compartment is NAPE localized? Again, we can provide only a very partial answer to this question. Experiments in which the membranes of intact neurons were probed with bacterial PLD (*Streptomyces chromofuscus*) suggest that a large portion of NAPE (at least 40%) is collected in the plasma membrane, where one would expect a precursor for an intercellular signaling molecule.[34]

Hydrolytic cleavage of membrane NAPE by a stimulus-activated PLD activity yields, in a single-step reaction, saturated and unsaturated NAEs. In fact, the NAEs recovered after stimulation of neurons in primary culture closely correspond to those expected from the structure of neuronal NAPE. This is an important piece of evidence in favor of a role of NAPE as NAE precursor but not the only one. Other supportive evidences include the ability of homogenates of neurons and other tissues (e.g., testis) to hydrolyze radiolabeled NAPE forming NAEs and the demonstration that NAPE turnover accompanies NAE formation in stimulated neurons.[29,35]

To understand the mechanism of NAE formation we need to learn more about the PLD activity which has been hypothesized to catalyze this reaction. This activity has not yet been purified, and indeed even its enzymatic characterization is still very partial and unsatisfactory. But as other PLD isoforms appear finally to yield to molecular cloning, we can hope that this avenue of experimentation will allow us to circumvent the difficulties which are likely to be encountered in the purification of such an enzyme.

ANANDAMIDE INACTIVATION: UPTAKE AND ENZYMATIC HYDROLYSIS

As a rule, the physiological actions of neurotransmitter substances, in the central nervous system as in the periphery, are terminated via either of two distinct mechanism of inactivation, hydrolytic degradation or cellular uptake. For instance, acetylcholine released from cholinergic neurons in the synaptic cleft is rapidly hydrolyzed to choline and acetate by the action of acetylcholinesterase, an enzyme which is concentrated on the surface of cholinoceptive cells. Other neurotransmitters, for example, the biogenic amines, are cleared from the synaptic sites by energy-dependent transporter proteins and are degraded within the presynaptic terminals. In the case of anandamide, it appears that both uptake and hydrolytic degradation may play a role, although the relative importance of these processes in vivo remains to be established.

When added to the medium of neurons or astrocytes in culture, radioactive anandamide is rapidly taken up by the cells: this process occurs with the rapid time-course, high affinity, temperature-dependence, saturability and selectivity typical of a carrier-mediated uptake[29] (Fig. 6.11). Further experiments will have to address the precise mechanism and kinetic properties of this uptake system in neurons and astrocytes, in order to determine which of these cell types takes up anandamide more effectively. Also, it will be essential to establish whether anandamide uptake is present in the adult central nervous system and whether it is localized to synaptic fields where anandamide release is expected to take place (e.g., where cannabinoid receptors are localized).

In addition to uptake, anandamide may be disposed of by a process of enzymatic hydrolysis. This reaction yields arachidonate and ethanolamine (Fig. 6.11), and it is carried out by a membrane-associated amidohydrolase activity which has been characterized kinetically in some detail and has been partially purified from porcine brain.[36-38]

In rat brain microsomes, anandamide amidohydrolase activity is optimal at pH 6 and 8, is independent of divalent cations, has an apparent K_M for [^{3}H]anandamide of about 13 μM and has a V_{max} of about 5600 pmol/minutes/mg protein. Its activity is blocked by both serine protease inhibitors (e.g., phenyl methyl sulphonylfluoride) and histidine alkylating reagents (e.g., *p*-bromophenacylbromide). By contrast, non-selective peptidase inhibitors (bacitracin, *o*-phenantroline) have no effect.[36]

Small changes in the chemical structure of anandamide are important for the catalytic activity of anandamide amidohydrolase (Fig. 6.12). Structural modifications that result in reduced hydrolysis include: (1) elongating the fatty acyl chain above 20 carbon atoms; (2) replacing *cis* double bonds with *trans* double bonds; (3) decreasing the number of double bounds to one, or eliminating them.[36]

It should be emphasized that, while rat brain microsome amidohydrolase shows little or no activity towards N-palmitoylethanolamine and other saturated NAEs, another enzyme activity

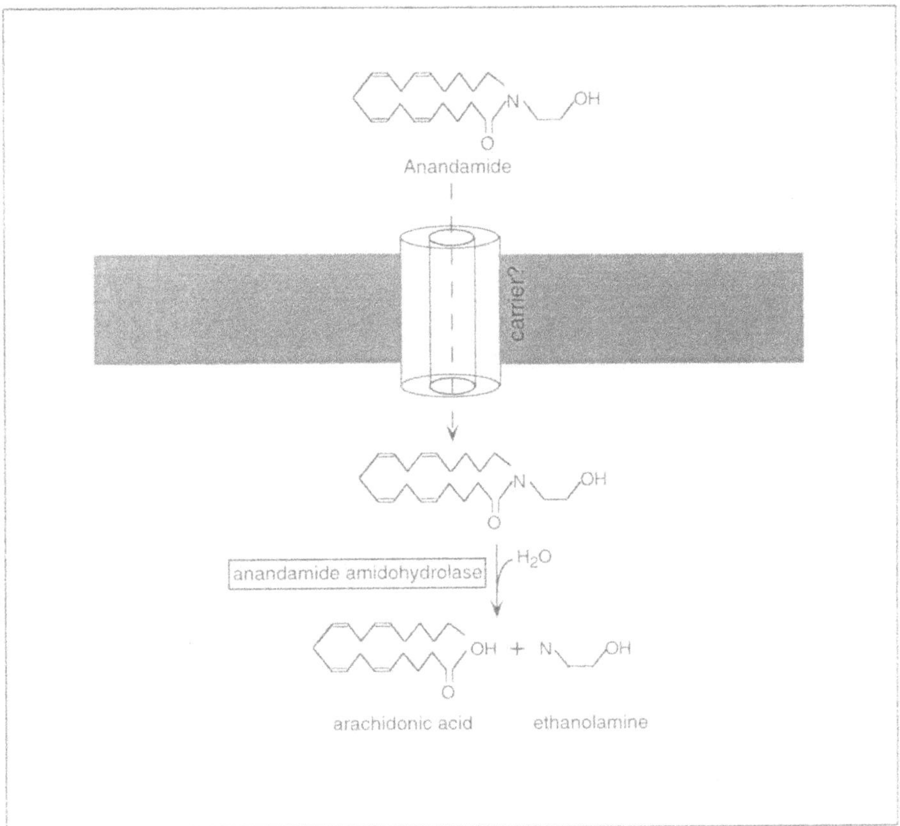

Fig. 6.11. Mechanisms of anandamide inactivation. For clarity, the anandamide amidohydrolase reaction is shown to occur intracellularly, but the actual localization of this enzyme is still unknown.

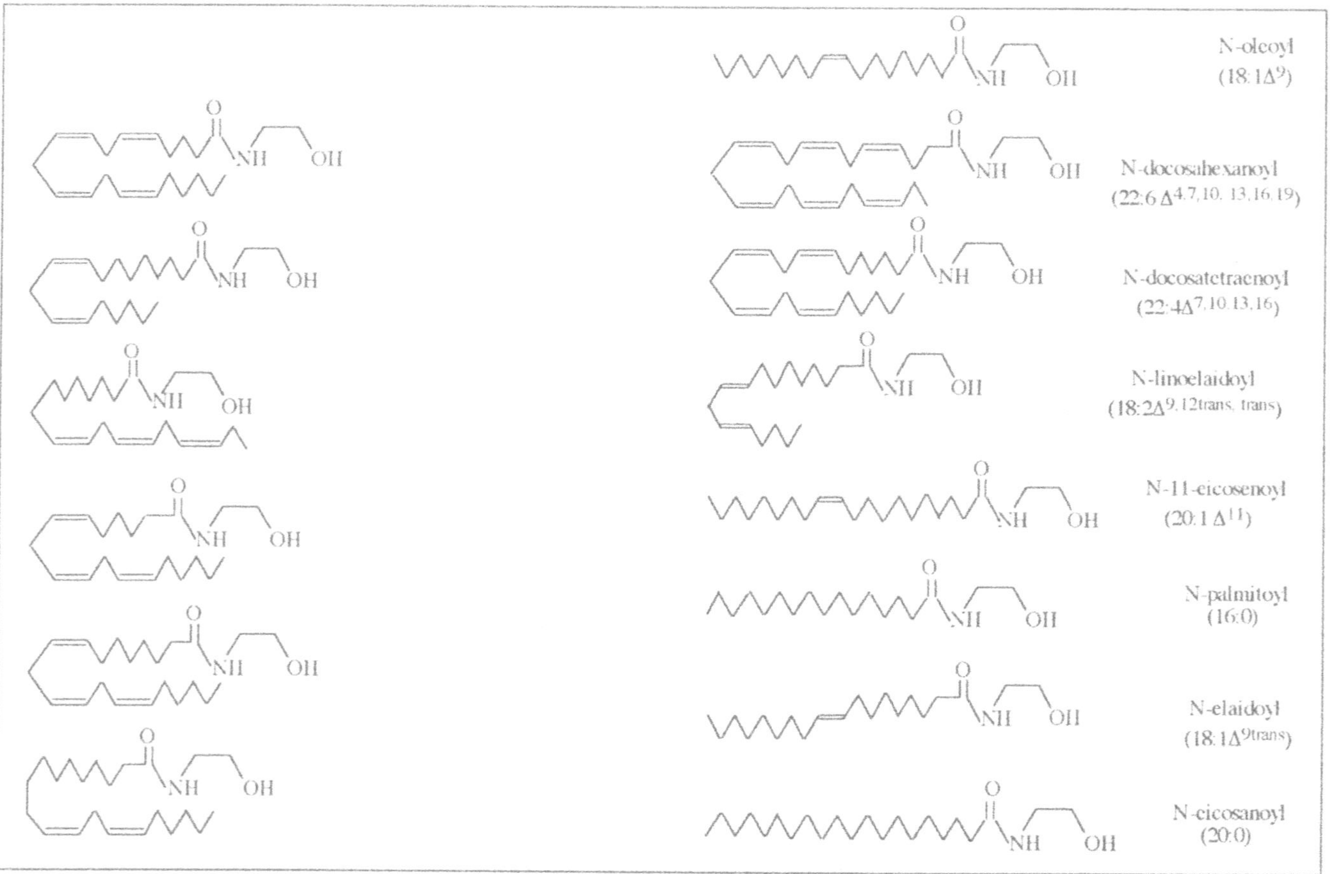

Fig. 6.12. Specificity of rat brain microsome anandamide amidohydrolase activity. Enzyme activity was measured by incubating the microsome fraction with [^{3}H]anandamide (14 μM) alone or in the presence of one of the indicated N-acylethanolamines (200 μM). (Reproduced with permission from Desarnaud F et al. J Biol Chem 1995; 270:6030-6035.)

in rat brain particulate fractions hydrolyzes N-palmitoyl-ethanolamine quite effectively, suggesting that a distinct amidohydrolase isoform may be implicated in the degradation of saturated NAEs.[36]

Amidohydrolase activity is concentrated in liver and brain tissues and, within the brain, in regions rich in cannabinoid receptors (e.g., globus pallidus and hippocampus) (Fig. 6.13). The parallel distribution of receptors and degradative enzyme in the central nervous system is particularly significant, as it supports a participation of anandamide amidohydrolase in the biological disposition of anandamide at its sites of action.

Finally, a feature of the anandamide amidohydrolase activity which deserves to be mentioned is its ability to carry out the reversal of anandamide hydrolysis. Under appropriate experimental conditions, the amidohydrolase is quite efficient in catalyzing the synthesis of anandamide from non-esterified arachidonate and ethanolamine. These conditions are far from being physiological, however, and include high concentrations of fatty acid and ethanolamine, an alkaline pH optimum and long incubation times.[38] As pointed out by Shozu Yamamoto and collaborators,[38] the reversibility of the anandamide amidohydrolase reaction may well account for the 'anandamide synthetase' activity in brain homogenates, as described by several groups.[39-41] In fact, the similar kinetic properties, pH optima and elution profiles (after several chromatographic fractionations) strongly support the notion that anandamide amidohydrolase and anandamide synthetase activities reside in a single enzyme.

LIPOXYGENASE METABOLISM OF ANANDAMIDE

The close structural resemblance of anandamide and arachidonic acid has prompted a series of biochemical experiments to test the possibility that anandamide serves as a substrate for enzymes that metabolize the polyunsaturated fatty acid. These experiments have considerable interest because, in principle, oxidative metabolism may modify the biological properties of anandamide—for example, change its affinity for cannabinoid receptors.

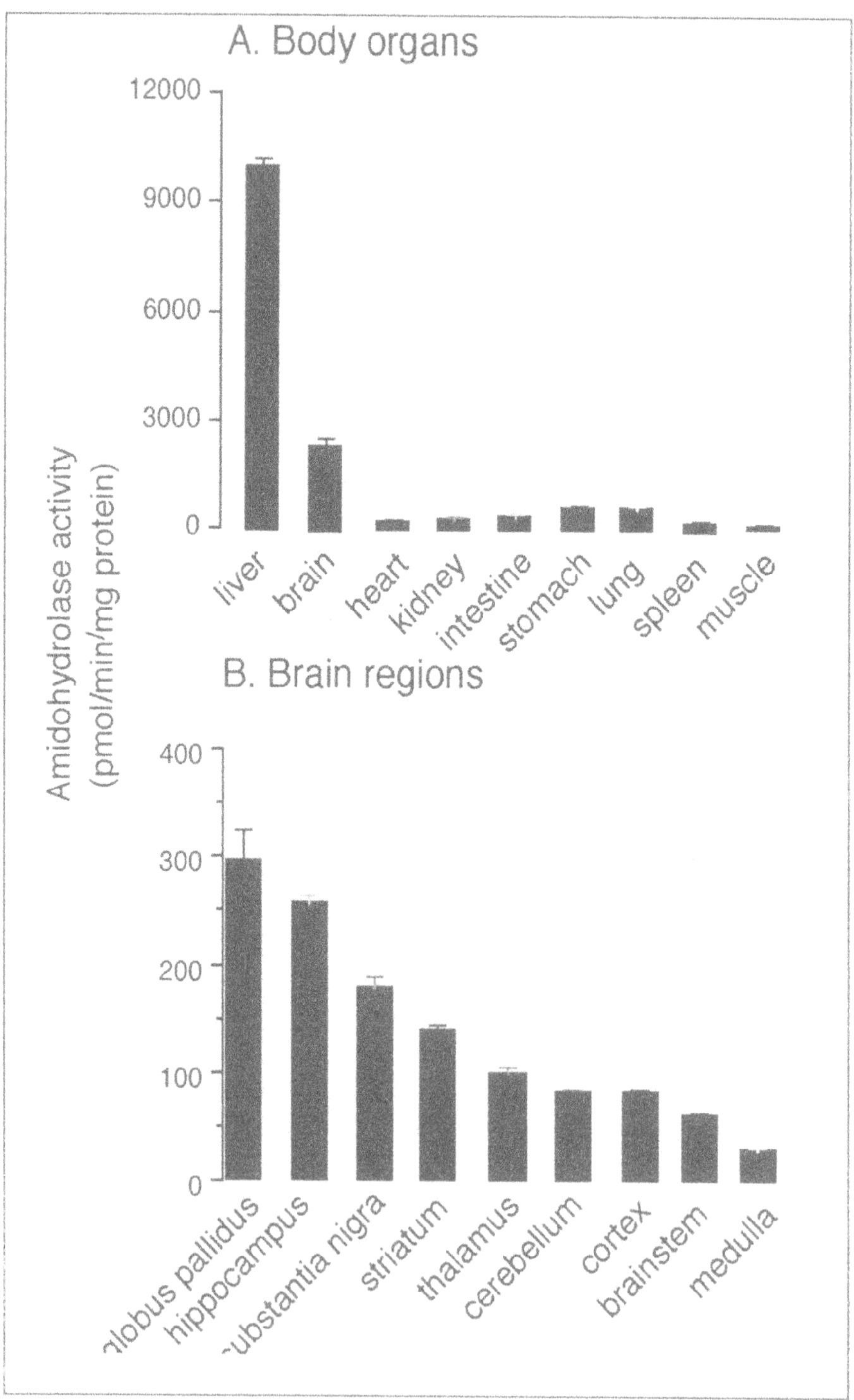

Fig. 6.13. Distribution of anandamide amidohydrolase activity in various organs (A) and various brain regions (B) of the rat. (Reproduced with permission from Desarnaud F et al. J Biol Chem 1995; 270:6030-6035.)

Unfortunately, the results obtained thus far have been disappointing. Anandamide was found to be, in vitro, a relatively good substrate for certain lipoxygenase enzymes—particularly 12-lipoxygenase.[42-43] But lipoxygenation does not appear strongly to modify its biological activity, as determined by a variety of functional assays used to screen cannabimimetic agents: at best, oxygenated derivatives are as potent as anandamide in binding to and activating cannabinoid receptors.[42-43] Moreover, and most importantly, no evidence has been reported yet for the formation of such lipoxygenase metabolites in situ. For lipoxygenation of anandamide to be considered more than a biochemical curiosity, future research will have to address this essential point.

A SECOND ENDOGENOUS CANNABINOID: *SN*-2-ARACHIDONOYL-GLYCEROL

In their 1992 *Science* paper, Mechoulam and coworkers mentioned that anandamide was only one of several lipid fractions containing cannabinoid-binding activity. They set out to characterize the additional active fractions and discovered that these were composed of either polyunsaturated NAEs very similar to anandamide (e.g., N-eicosatrienoylethanolamine)[44-45] or a distinct lipid component, *sn*-2-arachidonoyl-glycerol[46] (Fig. 6.14).

That polyunsaturated NAEs should mimic anandamide, to which they are structurally very similar, does not come as a great surprise. Moreover, the pharmacological properties of these NAEs—essentially indistinguishable from those of anandamide—and their scarcity in the brain relegate them, at least for the moment, to the role of anandamide's aliases. We cannot say the same of *sn*-2-arachidonoyl-glycerol. This lipid, considered until now a mere intermediate in glycerophospholipid turnover (see chapter 2), possesses one pharmacological property that makes it crucially different from anandamide: it binds to and it activates both CB1 and CB2 cannabinoid receptors with similar potencies (though it is less effective than anandamide at activating CB1 receptors).[46] As such, 2-arachidonoyl-glycerol is a possible candidate for the role of endogenous cannabinoid substance in peripheral tissues, where the predominant cannabinoid receptors are of the CB2 subtype.

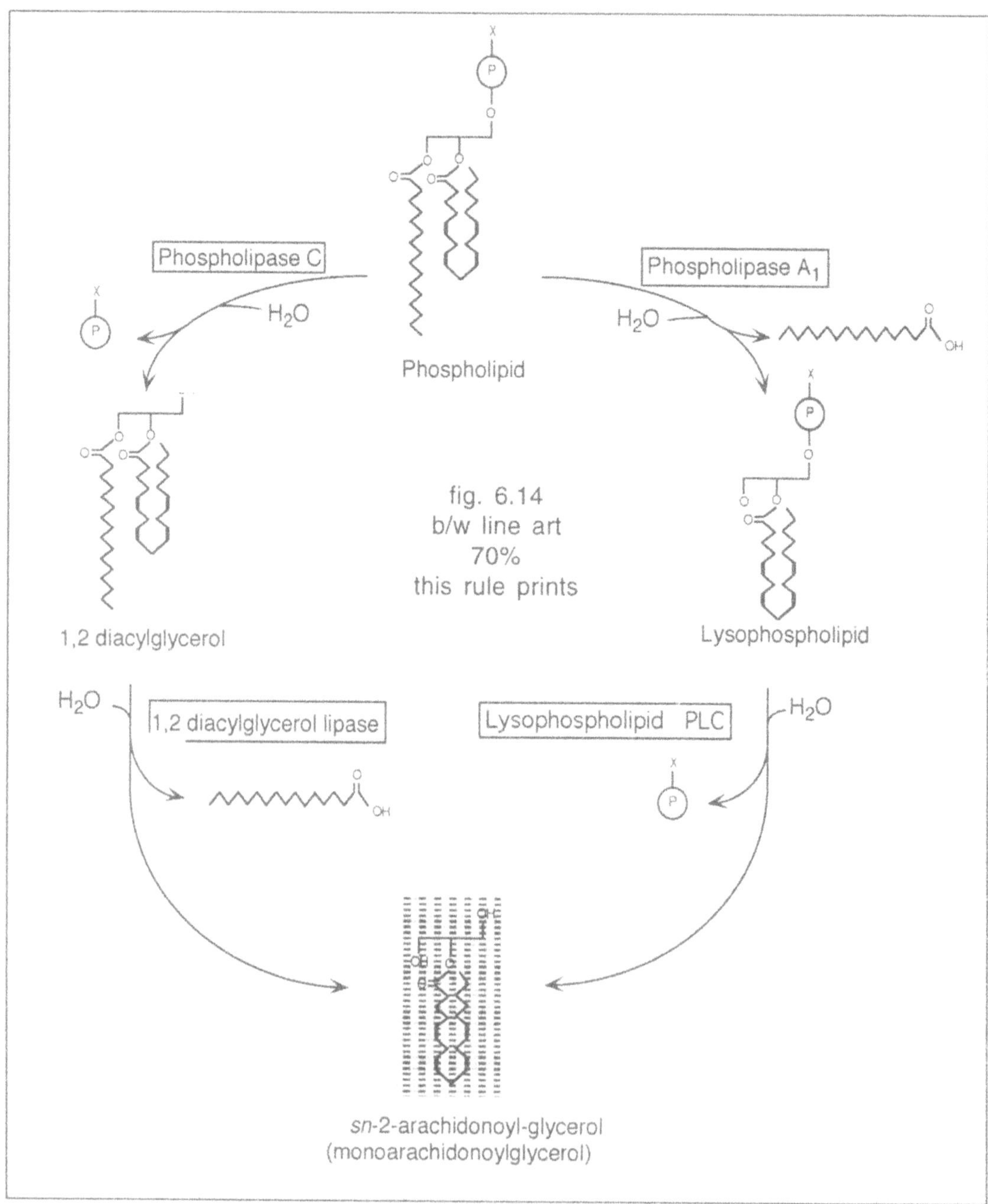

Fig. 6.14. Hypothetical pathways of sn-2-arachidonoylglycerol *formation in mammalian tissues. See text for details.*

The natural occurrence of *sn*-2-arachidonoyl-glycerol has been demonstrated in dog intestine[46] and evidence for its occurrence in rat brain has been reported.[47] Its biogenetic mechanism is still unknown, however. A plausible model, based on our understanding of phospholipid turnover, is shown in Figure 6.14. According to this scheme, *sn*-2-arachidonoyl-glycerol may derive from either the hydrolysis of 1,2 diacylglycerol, catalyzed by diacylglycerol lipase, or the hydrolysis of lysophospholipid, catalyzed by a lysophospholipid-specific PLC. Which, if either, of these pathways participates in *sn*-2-arachidonoyl-glycerol under physiological conditions remains to be established.

OPEN QUESTIONS

Perhaps more than any other area of the arachidonate field, research on anandamide begs for a conjunction of in situ biochemistry and physiology. We have learned much over the past three years on the behavioral effects of anandamide, but we still need to know whether this lipid mediator is produced physiologically in the adult brain: under what conditions, in which regions, and in what quantities.

Many important tools to carry out these studies are still lacking. For example, future work will have to tackle such difficult tasks as the purification and molecular cloning of the enzymes involved in the biogenesis of anandamide. This will lead the way to the localization of anandamide-producing cells in brain and in peripheral tissues.

We also need to learn more about the effects of anandamide on the electrical properties of central neurons. Thus far, the electrophysiological actions of anandamide have been studied for the most part using heterologous expression systems. These are of course useful pharmacological models, but they bear little resemblance to the functionally diverse neurons which constitute the adult mammalian central nervous system.

Finally, to understand the possible physiological functions of the endogenous cannabinoids—their roles in normal and pathological brain activity—pharmacological agents targeting the cascade of anandamide formation, release, uptake and degradation will have to be developed. Such drugs, which undoubtedly will be invaluable research tools, may also provide novel therapeutic approaches to diseases whose clinical, biochemical and pharmacological features suggest a link with the endogenous cannabinoid system.

REFERENCES

1. Devane WA, Hanus L, Breuer A., Pertwee RG, Stevenson LA, Griffin G, Gibson D, Mandelbaum A, Etinger A, Mechoulam R. Isolation and structure of a brain constituent that binds to the cannabinoid receptor. Science 1992; 258:1946-1949.
2. Mechoulam R. The pharmacohistory of cannabis sativa. In: Mechoulam R, ed. Cannabinoids as Therapeutic Agents. Boca Raton: CRC Press 1986; 1-16.
3. Piomelli D. Breve ma veridica storia della canapa indiana. Rome: Stampa Alternativa 1995 (in Italian).
4. Gaoni Y, Mechoulam R. Isolation, structure and partial synthesis of an active constituent of hashish. J Am Chem Soc 1964; 86:1646-1647.
5. Dewey WL. Cannabinoid pharmacology. Pharmacological Reviews 1986; 38:151-178.
6. Howlett AC, Qualy JM, Khachaturian LL. Involvement of G_i in the inhibition of adenylate cyclase by cannabimimetic drugs. J Pharmacol Experim Therap 1985; 29:307-313.
7. Howlett AC, Bidaut-Russel M, Devane WA, Melvin LS, Johnson MR, Herkenham M. The cannabinoid receptor: biochemical, anatomical and behavioral characterization. Trends Neurosci 1990; 13:420-423.
8. Matsuda LA, Lolait SJ, Brownstein MJ, Young AC, Bonner TI. Structure of a cannabinoid receptor and functional expression of the cloned cDNA. Nature 1990; 346:561-564.
9. Felder CC, Veluz JS, Williams HL, Briley EM, Matsuda LA. Cannabinoid agonists stimulate both receptor-dependent and non-receptor-mediated signal transduction pathways in cells transfected with and expressing cannabinoid receptor clones. Molecular Pharmacol 1992; 42:838-845.
10. Henry DJ, Chavkin C. Activation of inward-rectifying potassium channels (GIRK1) by co-expressed rat brain cannabinoid receptors in *Xenopus* oocytes. Neurosci Letters 1995; 186:91-94.
11. Mackie K, Hille B. Cannabinoid inhibit N-type calcium channels in neuroblastoma-glioma cells. Proc Natl Acad Sci USA 89: 3825-3829.

12. Mackie K, Lai Y, Westenbroek R, Mitchell R. Cannabinoids activate an inwardly-rectifying potassium conductance and inhibit Q-type calcium currents in AtT20 cells transfected with rat brain cannabinoid receptors. J Neurosci 1995; 15:6552-6561.
13. Galiègue S, Mary S, Marchand J, Dussossoy D, Carrière D, Carayon P, Bouaboula M, Shire D, Le Fur G, Casellas P. Expression of central and peripheral cannabinoid receptors in human immune tissues and leukocyte subpopulations. Eur J Biochem 1995; 232:54-61.
14. Matsuda LA, Bonner TI, Lolait SJ. Localization of cannabinoid receptor mRNA in rat brain. J Comp Neurol 1993; 327:535-550.
15. Herkenham M, Lynn AB, Johnson MR, Melvin LS, de Costa BR, Rice KR. Characterization and localization of cannabinoid receptors in rat brain: a quantitative in vitro autoradiographic study. J Neurosci 1991; 11:563-582.
16. Evans FJ. Cannabinoids: the separation of central from peripheral effects on a structural basis. Planta Medica 1991; 57:S60-S67.
17. Hollister LE. Health aspects of cannabis. Pharmacol Reviews 1986; 38:1-20.
18. Iversen LL. Medical uses of marijuana? Nature 1993; 365:12-13.
19. Munro S, Thomas KL, Abu-Shaar M. Molecular characterization of a peripheral receptors for cannabinoids. Nature 1993; 265:61-65.
20. Bayewitch M, Avidor-Reiss T, Levy R, Barg J, Mechoulam R, Vogel Z. The peripheral cannabinoid receptor: adenylate cyclase inhibition and G protein coupling. FEBS Letters 1995; 375:143-147.
21. Evans DM, Johnson RM, Howlett AC. Ca^{2+}-dependent release from rat brain of cannabinoid receptor binding activity. J Neurochem 1992; 58:780-782.
22. Evans DM, Lake JT, Johnson RM, Howlett AC. Endogenous cannabinoid receptor binding activity released from rat brain slices by depolarization. J Pharmacol Experim Therap 1994; 268: 1271-1277.
23. Mechoulam R, Hanus L, Martin BR. Search for endogenous ligands of the cannabinoid receptor. Biochem Pharmacol 1994; 48:1537-1544.
24. Fride E, Mechoulam R. Pharmacological activity of the cannabinoid receptor agonist, anandamide, a brain constituent. Eur J Pharmacol 1993; 231:313-314.
25. Facci L, Dal Toso R, Romanello S, Buriani A, Skaper SD, Leon A. Mast cells express a peripheral cannabinoid receptor with differential sensitivity to anandamide and palmitoylethanolamide. Proc Natl Acad Sci USA 1995; 92:3376-3380.
26. Venance L, Piomelli D, Glowinski J, Giaume C. Inhibition by anandamide of gap junctions and intercellular calcium signalling in striatal astrocytes. Nature 1995; 276:590-592.

27. Duffy S, Mac Vicar B. Adrenergic calcium signaling in astrocyte networks within the hippocampal slice. J Neurosci 1995; 15: 5535-5550.
28. Sontheimer H. Glial influences on neuronal signaling. The Neuroscientist 1995; 1:123-126.
29. Di Marzo V, Fontana A, Cadas H, Schinelli S, Cimino G, Schwartz JC, Piomelli D. Formation and inactivation of endogenous cannabinoid anandamide in central neurons. Nature 1994; 372:686-691.
30. Bachur NR, Masek K, Melmon KL, Udenfriend S. Fatty acid amides of ethanolamine in mammalian tissues. J Biol Chem 1965; 240:1019-1024.
31. Schmid HHO, Schmid PC, Natarajan V. N-acylated glycerophospholipids and their derivatives. Prog Lipid Res 1990; 29:1-43.
32. Natarajan V, Schmid PC, Reddy PV, Schmid HHO. Catabolism of N-acyl-ethanolamine phospholipids by dog brain preparations. J Neurochem 1984; 42:1613-1619.
33a. Natarajan V, Schmid PC, Reddy PV, Zuzarte-Agustin ML, Schmid HHO. Biosynthesis of N-acyl-ethanolamine phospholipids by dog brain preparations. J Neurochem 1983; 41:1303-1312.
33b. Cadas H, Gaillet S, Beltramo M, Venance L, Piomelli D. Biosynthesis of an endogenous cannabinoid precursor in neurons and its control by calcium and cAMP. J Neurosci 1996; 16:3934-3942.
34. Cadas H, Schinelli S, Piomelli D. Membrane localization of N-acylphosphatidylethanolamine in central neurons: studies with exogenous phospholipases. J Lipid Mediators 1996; in press.
35. Sugiura T, Kondo S, Sukagawa A, Tonegawa T, Nakane S, Yamashita A, Waku Z. Enzymatic synthesis of anandamide, an endogenous cannabinoid receptor ligand, through N-acylphosphatidylethanolamine pathway in testis: involvement of Ca^{2+}-dependent transacylase and phosphodiesterase activities. Biochem Biophys Res Commun 1996; 218:113-117.
36. Desarnaud F, Cadas H, Piomelli D. Anandamide amidohydrolase activity in rat brain microsomes. Identification and partial characterization. J Biol Chem 1995; 270:6030-6035.
37. Hillard C, Wilkinson DM, Edgemond WS, Campbell WB. Characterization of the kinetics and distribution of N-arachidonoylethanolamine (anandamide) hydrolysis by rat brain. Biochim Biophys Acta 1995; 1257:249-256.
38. Ueda N, Kurahashi Y, Yamamoto S, Tokunaga T. Partial purification and characterization of the porcine brain enzyme hydrolyzing and synthesizing anandamide. J Biol Chem 1995; 270:23823-23827.

39. Deutsch DG, Chin SA. Enzymatic synthesis and degradation of anandamide, a cannabinoid receptor agonist. Biochem Pharmacol 1993; 46:791-796.
40. Devane WA, Axelrod J. Enzymatic synthesis of anandamide, an endogenous ligand for the cannabinoid receptor, by brain membranes. Proc Natl Acad Sci USA 1994; 91:6698-6701.
41. Kruszka KK, Gross RW. The ATP- and CoA-independent synthesis of arachidonoylethanolamide. J Biol Chem 1995; 269: 14345-143348.
42. Ueda N et al. Lipoxygenase-catalyzed oxygenation of arachidonoylethanolamide, a cannabinoir receptor agonist. Biochim Biophys Acta 1995; 1254:127-134.
43. Hampson AJ, Hill WAG, Zan-Phillips M, Makriyannis A, Leung E, Eglen RM, Bornheim LM. Anandamide hydroxylation by brain lipoxygenase: metabolite structures and potencies at the cannabinoid receptor. Biochim Biophys Acta 1995; 1259:173-179.
44. Hanus L, Gopher A, Almog S, Mechoulam R. Two new unsaturated fatty acid ethanolamides in brain that bind to the cannabinoid receptor. J Med Chem 1993; 36:3032-3034.
45. Pertwee R, Griffin G, Hanus L, Mechoulam R. Effects of of two endogenous fatty acid ethanolamides on mouse vas deferentia. Eur J Pharmacol 1994; 259:115-120.
46. Mechoulam R, et al. Identification of an endogenous 2-monoglyceride, present in canine gut, that binds to cannabinoid receptors. Biochem Pharmacol 1995; 50:83-90.
47. Sugiura T, Kondo S, Sukagawa A, Nakane S, Shinoda A, Itoh K, Yamashita A, Waku Z. 2-Arachidonoylglycerol: a possible endogenous cannabinoid receptor ligand in brain. Biochem Biophys Res Commun 1995; 215:89-97.

INDEX

www.ingramcontent.com/pod-product-compliance
Ingram Content Group UK Ltd.
Pitfield, Milton Keynes, MK11 3LW, UK
UKHW051325070726
13610UKWH00014B/93

* 9 7 8 3 6 6 2 0 5 8 0 8 4 *